ROOF FRAMING

ROOF FRAMING

H. H. SIEGELE

Published in 1982 by
Sterling Publishing Co., Inc.
Two Park Avenue
New York, N.Y. 10016

Library of Congress Catalog Card Number: 74-25285

Siegle, H.H.
Roof Framing.
New York
Feb. 1975

ISBN: 0-8069-8626-3 Paper
8625-5 Library
Previously
ISBN: 87749-236-0 (Cloth)
ISBN: 87749-792-3 (Paper)

1986 Printing

Printed in the United States of America.

PREFACE

This Volume has been prepared so that every carpenter, and particularly every carpenter apprentice, will have available a book that gives in a practical way the fundamental things about roof framing. To present roof framing problems, so that the experienced roof framer will find in them something new about his work is not enough. A thorough treatment of roof framing must first of all start with the simple things that the inexperienced boy who wants to become a carpenter does not know. In order to do that effectively, the illustrations will have to be practical, which is to say, that the illustrator must be a man who has actually framed roofs, and knows just where the beginner is likely to need help, and give that help by means of text and illustrations. This does not mean that the student will find his problems all worked out—not at all—but he will find all of the important parts explained, so that he can go ahead and apply himself to his tasks as he gains experience, which will eventually make him a first class roof framer. When he has mastered everything that is in this book about roof framing, he will have accomplished an achievement that will be a lifetime asset to him, whether he follows carpentry or takes up some other kind of work.

The author is indebted to the Editor of THE CARPENTER for helpful suggestions in editing this work, when it appeared in the official journal of The United Brotherhood of Carpenters and Joiners of America.

THE AUTHOR

TABLE OF CONTENTS

TABLE OF CONTENTS

ROOF FRAMING

LESSON 1

DIFFERENT KINDS OF ROOFS

Apprentice and Journeyman.—It should be remembered, in connection with this study, that I shall keep uppermost in mind the needs of the apprentice carpenter; for it is he who needs help, rather than the journeyman carpenter. However, there are many journeymen carpenters who will find these lessons helpful to them; for roof framing is a job that some carpenters are called upon to do more frequently than others, and those who do not get much of it in the field, can find the help they need in this work. Many years ago when I was treating this subject, a journeyman carpenter in his fifties, remarked that "I learned more about roof framing from those

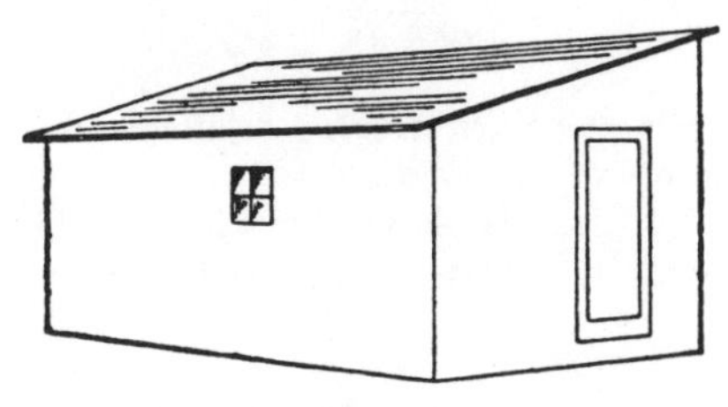

Fig. 1

articles, than I learned while working at the trade. . . ." This journeyman carpenter, perhaps, got little roof framing in his experience, because he lacked that first essential qualification, confidence in his ability to do the job without making damaging mistakes. On the other hand, the carpenter who had confidence in his ability, because he had framed other roofs and understood roof framing thoroughly, didn't trouble himself about reading the articles. This is logical, for why should one who understands a thing spend time that he can use for other purposes, studying something he already knows? But just because there are many carpenters who thoroughly understand roof framing is no reason why this branch of our trade should not be made clear to others who are lacking in that knowledge. Then there are those who feel

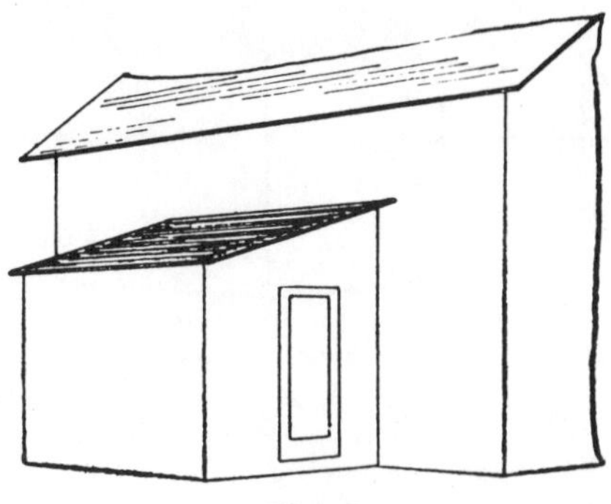

Fig. 2

that matters which are a sort of common knowledge among the average carpenters, should not be dealt with because "everybody knows that." But who is "everybody"? And how did this mythical individual come by his

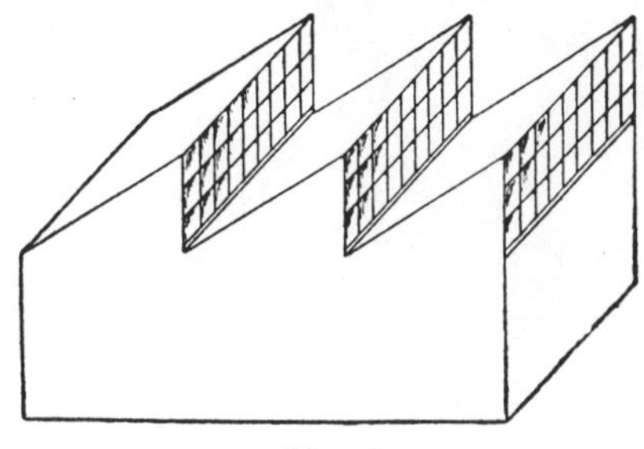

Fig. 3

knowledge? Isn't it a fact that sometime, somewhere in the past, he, himself, did not know everything he knows today? The answer is self-evident. In this work, the apprentice is presumed to be without the knowledge of many things until he has proved to the world that he understands them. And long before he understands everything about carpentry, will he be standing with the ranks of journeymen carpenters.

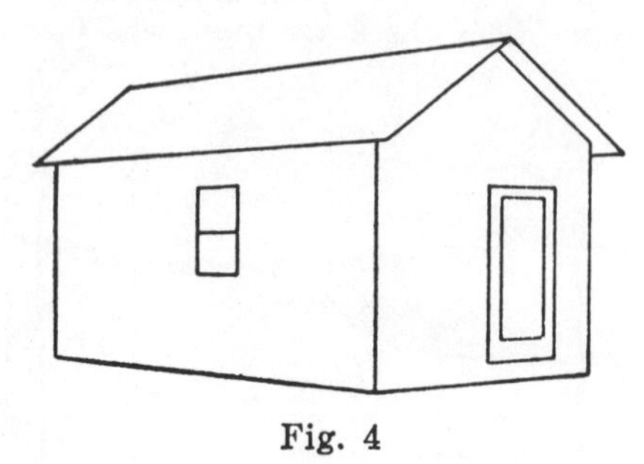

Fig. 4

Single-Pitch Roof.—Fig. 1 is a sort of diagram, showing a single-pitch roof. This roof is frequently used on temporary buildings, such as material sheds, stands and tool houses. When it is used on permanent buildings

Fig. 5

they are usually of the very cheapest kind of buildings. This roof is also called pent roof, shed roof, to-fall and, erroneously, lean-to.

Lean-To Roof.—Fig. 2 shows what is in reality a lean-to roof, which is to say, that the roof leans against some other building and derives some of its support from that building.

Saw-Tooth Roof.—The saw-tooth roof shown in Fig. 3, so far as the roof is concerned, is much on the order of a shed roof. This roof is mostly used on factory buildings, garages, and similar structures. The advantage of this roof is that it makes possible a great deal of window space for the admission of light and for ventilation.

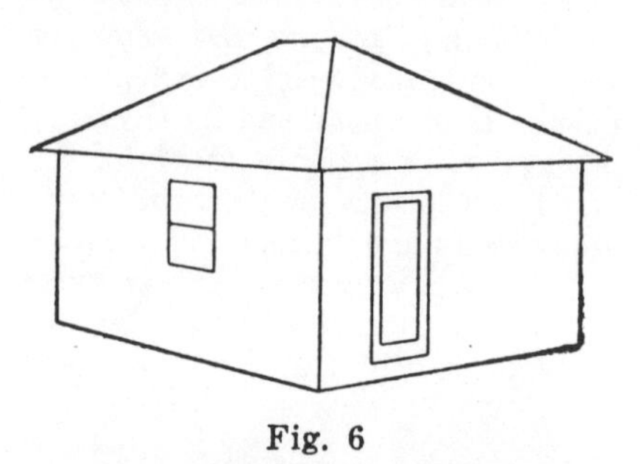

Fig. 6

Double-Pitch Roof.—A simple diagram of a building with a double-pitch roof is shown by Fig. 4. This roof is, perhaps, the basis for all

Fig. 7

other roofs that are classified as pitch roofs. It is also called gable roof.

M Roof.—An M roof is shown by Fig. 5, which is made up of two double-pitch roofs. This roof has two advantages; first, it reduces the elevation of the building, and second, much shorter material for rafters can be used in its construction. These advan-

Fig. 8

tages will be clear when one takes into consideration that the diagram we are showing represents a rather small building.

Hip Roof.—Fig. 6 shows a building with a hip roof. This roof is perhaps the strongest of the pitch roofs. It is much more substantial than ordinary conditions require; however, in localities where there is danger of wind damage, this roof is quite suitable,

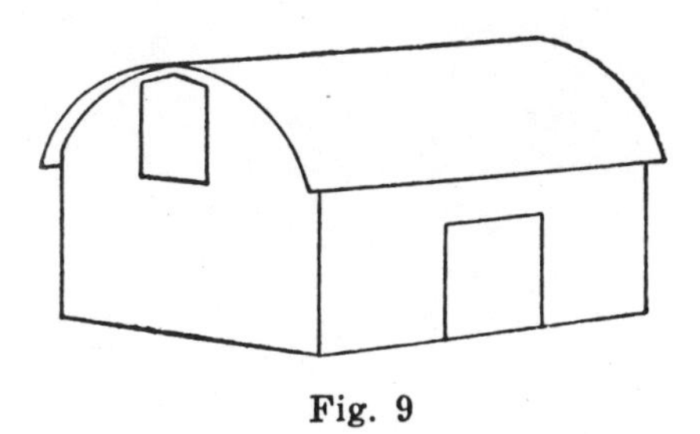

Fig. 9

but in such cases it must be firmly anchored to the building supporting it, which in turn must be anchored to the foundation. I recall an instance where a lean-to roof, which was a continuation of a double-pitch roof, was lifted by a strong wind and laid back onto the main roof, much as one would close a book.

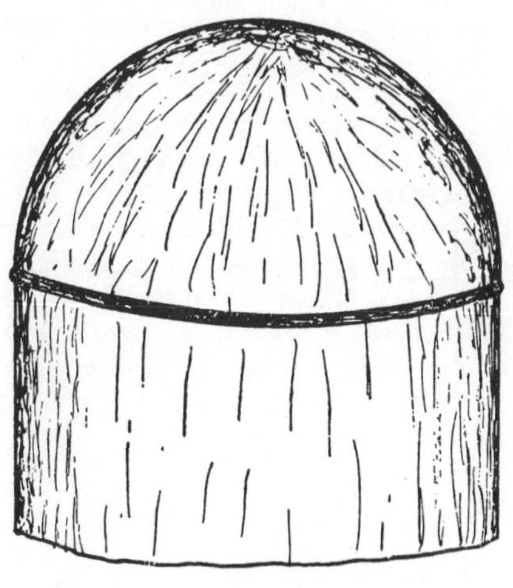

Fig. 10

Gambrel Roof.—A gambrel roof is shown by Fig. 7. The advantage of this roof is that it increases the attic space, and when dormer windows are used it is almost equivalent to a sec-

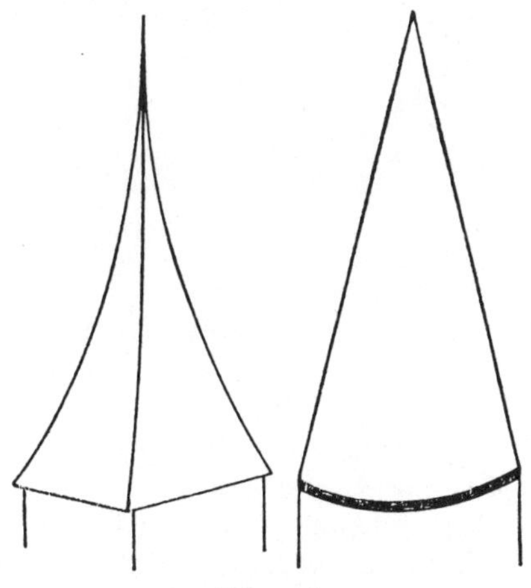

Fig. 11

ond story. A gambrel roof has little in its favor from the standpoint of appearance.

Mansard Roof.—Fig. 8 shows a

Mansard roof, which is a modification of the gambrel roof, or rather the hip-version of the gambrel roof. Its advantages lie in the space added to the attic and in the additional strength of the construction.

Semicircular Roof.—Fig. 9 shows a building with a semicircular roof. This roof is often used on barns. Sometimes a ridge is added to this by reversing the curvature slightly about four or five feet from the center at the top, which gives a sort of English Gothic effect. The curvature at the eaves is also reversed enough to give them a little more drip. See lesson 22.

Dome and Turret Roofs.—Fig. 10 shows a dome roof, while Fig. 11 shows to the left a square turret and to the right a cone-shaped turret.

Other Roofs.—Other roofs that might be added are: Bell roof, ogee roof, gable-and-valley roof, hip-and-valley roof, irregular-pitch or uneven-pitch roof, deck roof, and flat roofs. Most of the roofs shown or referred to here we shall directly or indirectly touch upon as we proceed with this work.

LESSON 2

ABC'S OF ROOF FRAMING

Up-To-Date Methods.—Two letters came to me in the same mail, in my early experience as a contributor to "The Carpenter" magazine. One of those letters was from an instructor

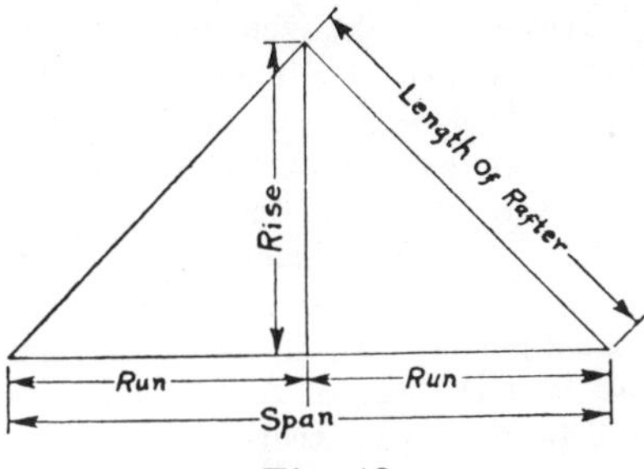

Fig. 12

in a vocational school and the other came from a carpenter, who evidently had had much experience in roof framing and, no doubt, understood roof framing thoroughly. Both men

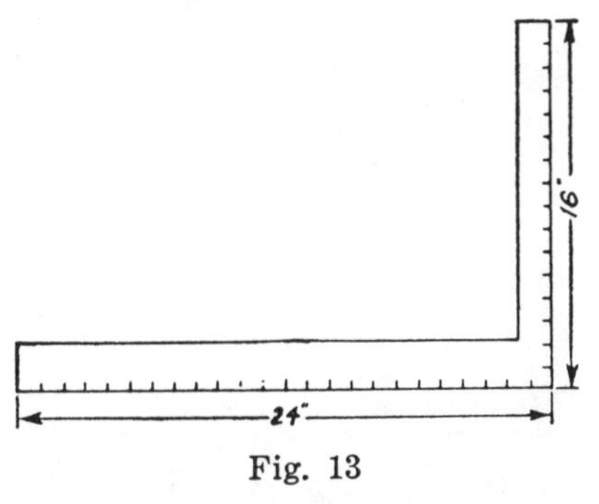

Fig. 13

condemned the use of the terms one-fourth pitch, one-third pitch, one-half pitch, and so forth, as being out of date, or as one of them put it, "hang-overs" from the past. I read the letters over and over to find what the up-to-date terms were that these men would use in place of the old-fashioned terms, but to my disappointment they did not furnish the substitutes. The nearest approach to furnishing such new terms was that they both advocated the use of "roof framing tables." Both men admitted that the apprentices should have an understanding of those old-fashioned things, but they should use up-to-date methods. I quote from one of the letters: "I

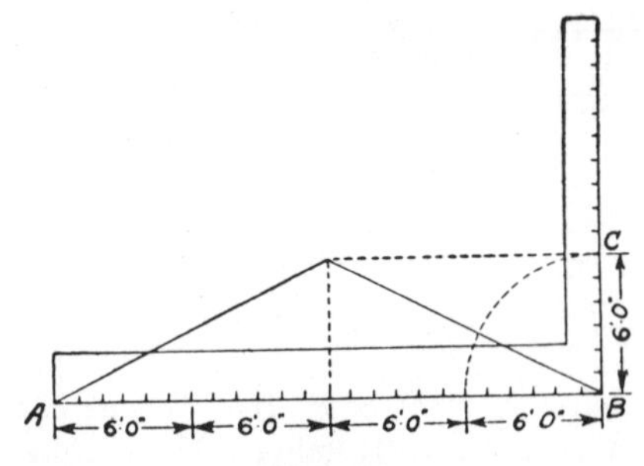

Fig. 14

think it is well to show all these things to the apprentices for their information, but the more accurate and up-to-date methods should be taught."

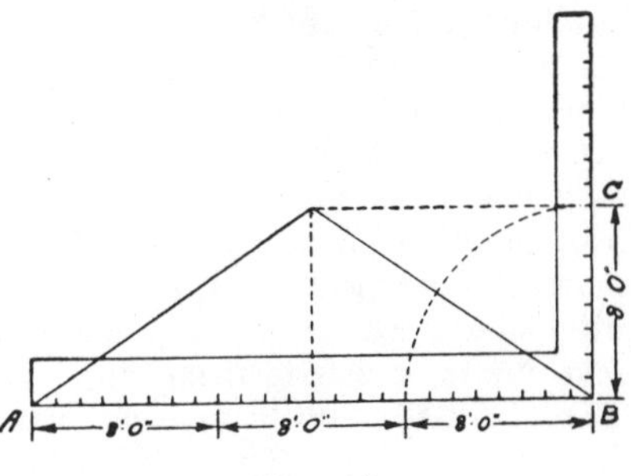

Fig. 15

Old-Fashioned Terms. — I have framed many roofs with my own hands and, while I have never used roof framing tables, I have examined and studied many of them, and it is my conclusion that roof framing tables do not furnish terms or expressions that can in any way measure up to the old-fashioned basic terms, one-fourth pitch, one-third pitch, one-half pitch, full pitch, and so on. These terms give the student a practical conception of the thing under consideration. But after these basic principles are mastered, the student can take up a roof framing table and discover that the table has everything figured out for him—that he can now go ahead and frame roofs without doing any thinking at all. But imagine the carpenter who should attempt to frame roofs without understanding anything about roof framing, excepting what the roof framing tables give him—imagine, I repeat, his embarrassment if perchance he can't find the table when he is ready to start framing a complicated roof with carpenters standing around waiting for orders. At such a time he would give anything to be able to frame any kind of roof by the simple application of the steel square. The steel-square method might be old-fashioned, but if thoroughly understood and carefully applied it will work every time.

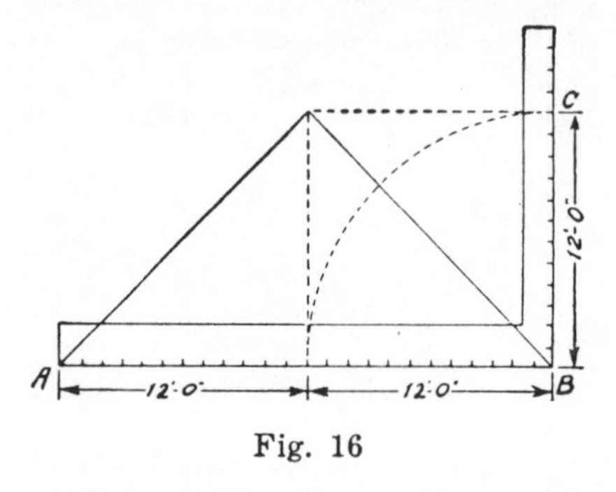

Fig. 16

ABC's of Roof Framing.—The illustrations of this lesson represent what I like to call the ABC's of roof framing, which are to roof framing what the alphabet is to the written language.

Span, Run, Rise and Rafter.—The diagram shown in Fig. 12 gives the first requisites of roof framing; namely, the span, the run, the rise, and the length of the rafter. These are all basic principles and they remain the same whether you use the terms I am using or express them by some other means—they are fundamental.

Steel Square.—Fig. 13 shows a steel square with a 24-inch blade and a 16-inch tongue. The blade of the steel square represents the basic span in roof framing, which is to say that the span for a 12-inch run is 24 inches. This principle does not change, so far as the common rafter is concerned. The rise is basic in name only, for the figures to be used for it change whenever there is a change in the pitch of the roof. This is further illustrated by the following diagrams:

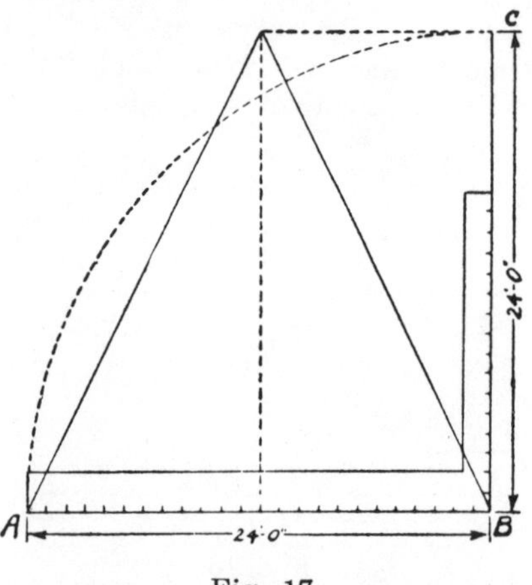

Fig. 17

One-Fourth Pitch.—Fig. 14 shows the steel square and its relationship to the span, run and rise of a one-fourth pitch roof. As shown, the inches of the square represent feet in the diagram. The span has been divided into four equal parts of 6 feet, and since the rise of a one-fourth pitch roof equals one-fourth of the span, we have projected one-fourth of the span from the blade of the square to the tongue, as shown by the dotted part-circle. The point where this part-circle intersects with the outside edge of the tongue has been carried to the left by dotted line un-

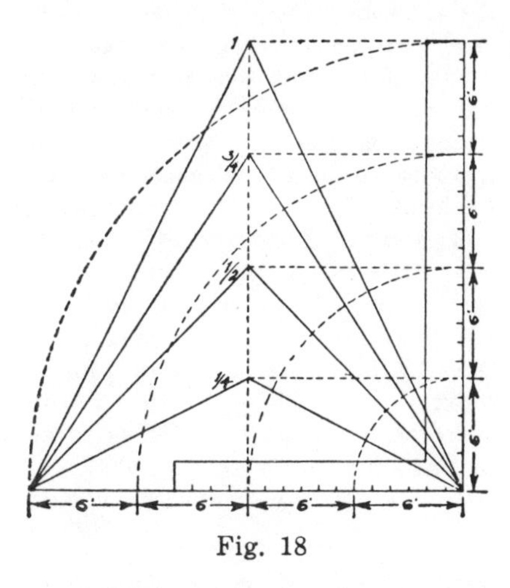

Fig. 18

til it intersected with the perpendicular dotted line representing the center. Now the rafter lines are drawn and the diagram is complete. The distance from *A* to *B* represents the span, while the distance from *B* to *C* gives the rise.

Other Pitches.—Fig. 15 shows a diagram of a one-third pitch roof. The explanations are the same as those given for the one-fourth pitch, excepting that the span has been divided into three equal parts of 8 feet, because the rise of a one-third pitch represents one-third of the span. Fig. 16 shows the span divided into two equal parts, since the diagram represents a one-half pitch roof, while in Fig. 17 the span is undivided, which represents a full pitch roof.

Basic Pitches.—Fig. 18 is a diagram showing four basic pitches brought together for comparison; they are one-fourth, one-half, three-fourths and full. I call these basic pitches because they are easily expressed and the figures to be used on the square do not involve fractions. It should be remembered that whenever there is a change in the rise, no matter how small it might be, it gives a different pitch. The run, though, remains basic, speaking of common rafters.

Full Pitch and Steeper.—The four basic pitches shown in Fig. 19, were used a great deal for steeples, up to a generation or so ago. These pitches are, reading from the bottom up, full

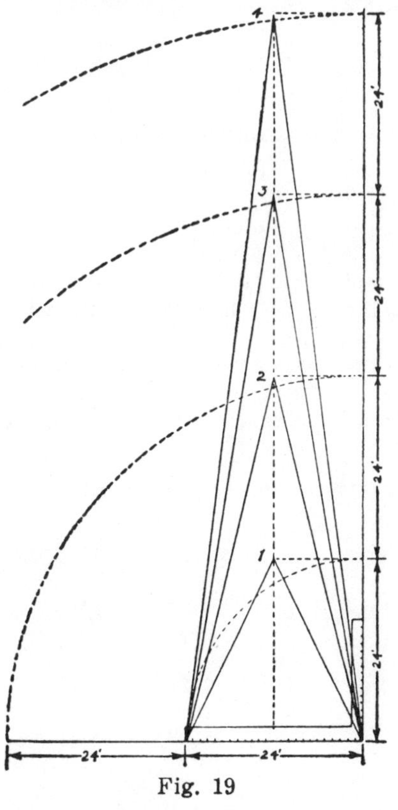

Fig. 19

pitch, double pitch, triple pitch and quadruple pitch. The student can readily see how this process can be carried out indefinitely, but pitches steeper than the quadruple pitch could hardly be called practical. Moreover, the number of pitches that are possible between any two basic pitches are innumerable.

LESSON 3

STEPPING OFF RAFTER PATTERNS

Indispensable Tool.—Old-fashioned as it is, the steel square is indispensable when it comes to roof framing. Even those roof framers who have discarded everything that is not up-to-the-minute, find it necessary to use this old-fashioned tool. But after analyzing the so-called new methods, I find that they are not new at all. When I was in common school, I was tempted to (and perhaps sometimes did) use those new methods in arithmetic, especially when I came to problems I couldn't master. How simple the study of arithmetic would have been, if instead of having to work out the problems, I could have turned to the back of the book and copied the answers—I could have known before going to class with certainty that I had all the answers right. The only difficulty I found with that method, back in my school days, was that if perchance I would have to demonstrate at the blackboard in detail how the answers were arrived at, I would have found myself out on that proverbial limb. And that is just where the rub will come in for the roof framer who uses the answers to roof framing problems without knowing for himself how to solve the problems.

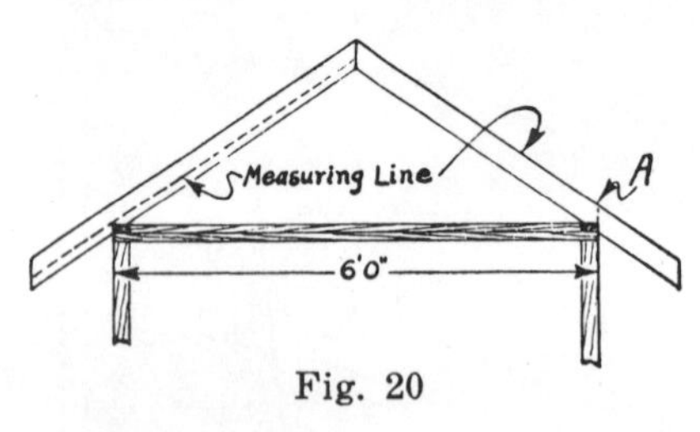

Fig. 20

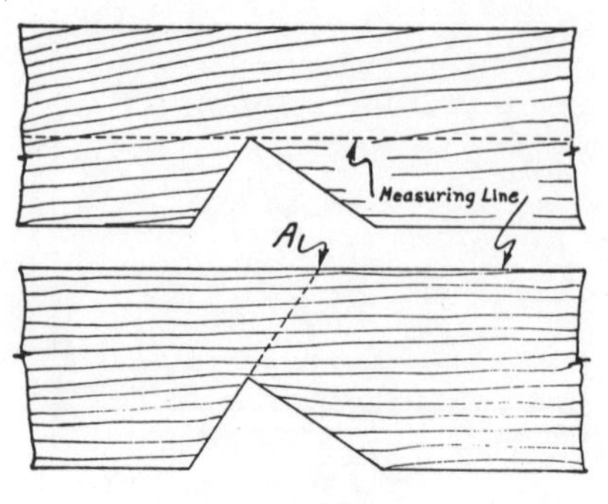

Fig. 21

A Method That Works.—In this work it is my purpose to give every apprentice a chance to learn for himself how to solve roof framing problems by means of the steel square. After that it will be safe for him to use other methods if he cares to do so. Whether the methods I shall use are old, or whether they are new, does not matter to me. The one thing I shall endeavor to avoid, is that of giving the answers to problems without showing how such answers are arrived at. That is my method of teaching—whether this is a new or an old method, I do not know—all I can say for it is that it will work—that the man who understands it thoroughly and applies it carefully will get results.

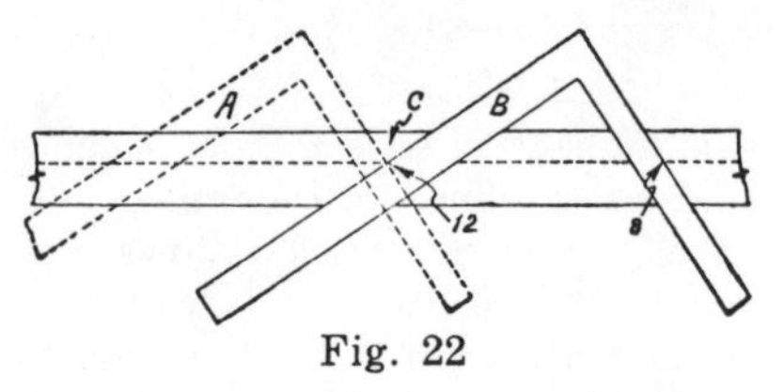

Fig. 22

Measuring Line.—Fig. 20 shows a pair of rafters in place for a roof having a 6-foot span. The dotted line shown on the rafter to the left is called, measuring line. The measuring line was used exclusively back in the days when rafter timbers were hewed into shape with a broad ax. But for a generation or two the measuring line has been used less and less, excepting in certain localities where what is known as "native stuff" is used for rafters. This material, due to seasoning or for some other reasons, is seldom perfectly straight. In such cases the offset measuring line will insure accuracy, whereas the edge of the timber would not. Perhaps the best

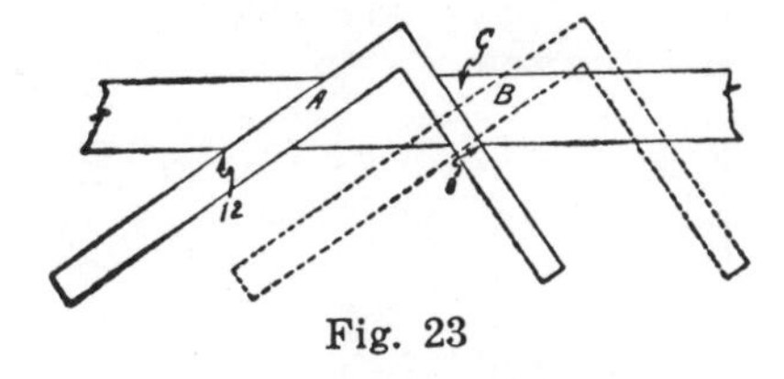

Fig. 23

way to make the measuring line is by striking a chalkline at the proper place on the pattern material from end to end. The general line of the timber must be straight, even though the edge might have small crooks and uneven places.

Edge Measuring Line.—At the present time, when rafter material comes to the job smooth on four sides and perfectly straight, the edge of the timber is usually used for the measuring line in stepping off rafter pat-

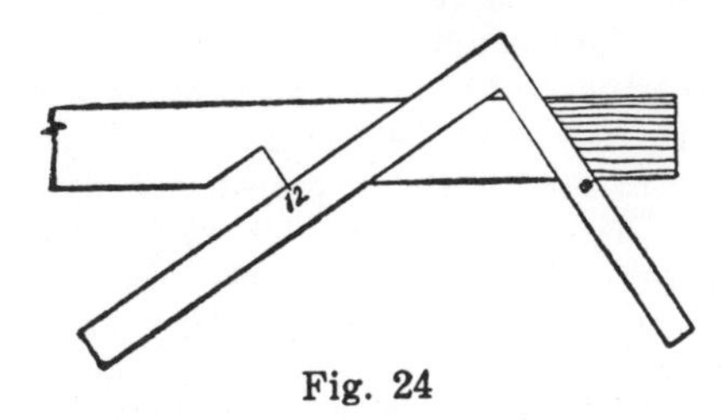

Fig. 24

terns. This is pointed out on the rafter to the right in Fig. 20. At *A* by dotted line we aıe showing how the plumb cut of the seat must be projected to the upper edge of the timber to establish the starting point for stepping off the pattern.

Fig. 21 further illustrates the two measuring lines. The upper drawing shows the relationship of the offset measuring line with the seat of the rafter, while the bottom drawing shows at *A*, how the plumb cut is projected to the edge of the material. The notch, which constitutes the seat of the rafter, is also called, bird's mouth. This term, though, belongs to the davs when round timbers were used for rafters, hewed on one side to make an even bearing for the sheeting. Then when the notch for the seat was cut, it resembled the wide-open mouth of a young bird begging for food, hence the name, bird's mouth.

Application of Square. — Fig. 22 shows the application of the steel square for marking the bird's mouth

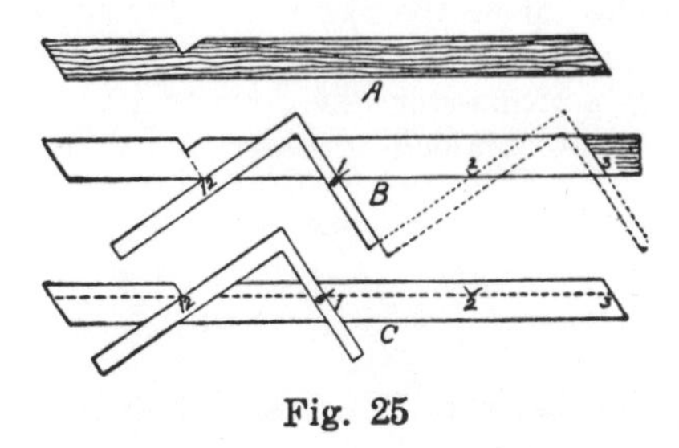

Fig. 25

when the offset measuring line is used. At *A*, by dotted lines, the square is shown in position for marking the plumb cut, while at *B* it is in the position for marking the horizontal cut. The figures used on the square are for a one-third pitch, 12 and 8, and as pointed out, these figures intersect with the measuring line. The part to be cut out is marked *C*.

How to apply the square when the edge of the timber is used as measuring line is shown in Fig. 23. Here the square marked *A* is in position for marking the plumb cut, while the square marked *B* gives the horizontal cut. The part to be cut out is again marked *C*. The figures to be used on the square are the same as in the other case, excepting that they intersect with the edge of the timber.

Marking Tail Cut.—The square in position for marking the tail cut of a rafter is shown in Fig. 24. The same figures, 12 and 8 are used, which intersect with the edge of the timber. The shaded part to the right is to be cut off.

Stepping Off.—At *A*, Fig. 25, we show a rafter for a 6-foot span cut and ready for use. At *B* we show the position of the square for the first step in stepping off rafter patterns, using the edge of the timber for measuring line. The square shown by dotted lines to the right gives the position for marking the comb cut. The part shown shaded is cut off. The steps are numbered 1, 2 and 3. At *C* we show the same rafter with an offset measuring line and the square in position for the first step. The points obtained in stepping off this

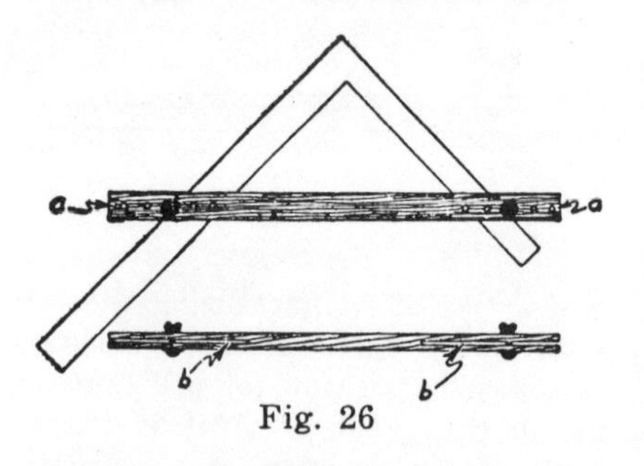

Fig. 26

rafter are marked, as in the other case, 1, 2 and 3. We will have more to say about stepping off rafter patterns in the next lesson.

Fence and Guides.—Fig. 26 shows two views of a job-made fence for a roof framer's square. The holes pointed out at *a, a* receive the thumbscrew bolts with which the fence is clamped to the square. At *b, b* are pointed out the slots into which the square is slipped. This fence gives satisfactory results if the pattern material is perfectly straight. But in cases where small humps are on the

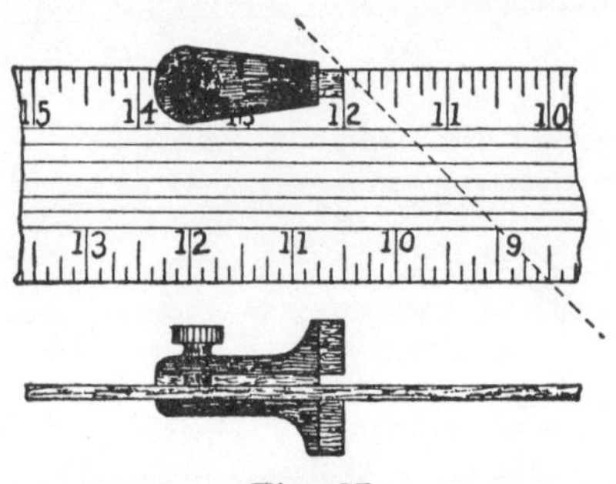

Fig. 27

edge of the timber the fence will cause the square to rock, which renders the marking unreliable. There are available on the market, guides that are free from the objection just pointed out. These giudes are made in pairs, much on the order of what is shown in Fig. 27. The upper drawing shows a side view, while the bottom drawing shows an edge view. The dotted line shows the relationship of the guides to the timber when they are set at 12 and 12 on the square.

In my early experience I used guides for stepping off rafter patterns, but now I simply hold the square in such a manner that the index fingers will serve as guides. I like this better than using guides, for guides limit the use of the square to stepping off.

LESSON 4

OBTAINING LENGTH OF RAFTERS

There are a number of ways to find the length of rafters in roof framing. A few of the most practical ones I shall explain in this lesson. The thing I want to avoid, as stated in the previous lesson, is to give the answers to roof framing problems without showing how such answers were found, which means that I am not giving any roof framing tables.

Roof Framing Tables.—I do not object to the use of roof framing tables—they have their proper place in the building industry. Architects, contractors, superintendents, estimators and engineers, whose work is mostly confined to offices, should use roof framing tables because to them, they are labor-saving devices. A man working in an office, who must know the

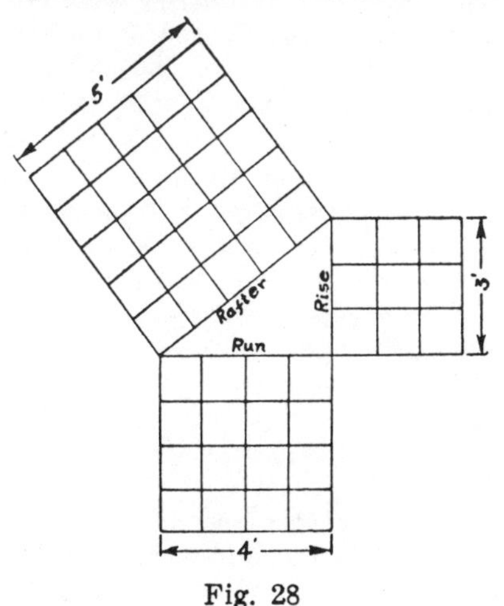

Fig. 28

lengths of roof material could hardly be expected to take a steel square and by that means obtain the different lengths of rafters that he must know. In fact, those men would find daily use for a reliable roof framing table, just as the carpenter finds daily use for his steel square. That being true, there would be as little likelihood of the roof framing table being misplaced in an estimator's office, as there would be for a carpenter to misplace his steel square. In other words, a carpenter who might not be called upon to frame a roof for years, could

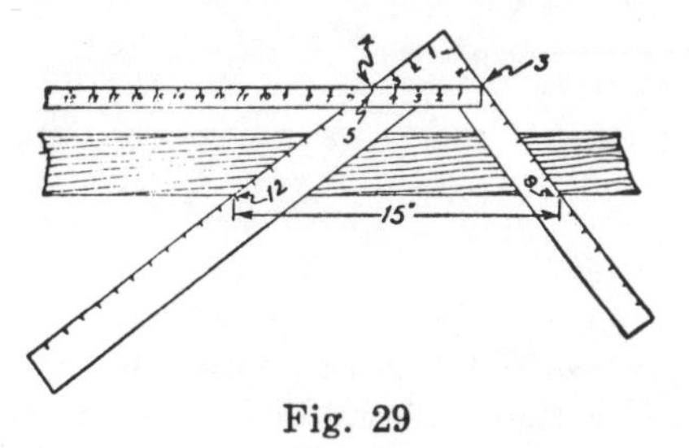

Fig. 29

easily lose or misplace a roof framing table, but because he uses his square daily he would hardly misplace it, and if he should lose it, it would be necessary for him to get another square almost immediately. Therefore I advocate that every apprentice should learn how to solve roof framing problems by means of the steel square.

Most Accurate Method.—The most accurate, but the least practical, method of finding the lengths of rafters is the square root method, which I am illustrating in Fig. 28. The diagram gives a run of 4 feet, a rise of 3 feet and a rafter of 5 feet. The length of the rafter was obtained by that old rule: "The base squared plus the altitude squared equals the hypotenuse squared." In other words, the square root of the run squared plus the rise squared equals the length of the rafter. I have purposely taken small figures for this illustration, in order to make the extraction of the square root simple. Let us work the problem: The run of 4 feet squared equals 16, and the rise of 3 feet squared equals 9—16 plus 9 equals 25. The square root of 25 must be greater than either the rise or the run, so we will try the next higher figure, 5—5 times 5 equals exactly 25, therefore 5 is the square root of the run squared plus the rise squared, or the length of the rafter in feet. This problem is simple, but suppose you had a run of 12 feet 6½ inches, and a rise of 8 feet 7¾ inches—that would complicate matters because it would involve a great deal more work, especially, since to be sure no mistake has been made, the whole procedure would have to be checked and proved.

Diagonal Method.—Fig. 29 shows another method of obtaining the length of rafter patterns. Here the inches on the square are taken as equaling feet. The rule is shown in a position to measure the diagonal distance between the figure 4 on the body of the square, and the figure 3 on the tongue. The distance between those two figures is 5 inches, which, if taken as feet, would be the same as we have found in the square root method. The figures 12 and 9, as shown on the square, are relatively the same, so far as the pitch is concerned, as 4 and 3. The diagonal distance between 12 and 9 is 15 inches, or the length of the rafter per foot run.

Stepping Off Method. — The most practical, and if carefully done, so accurate that it will serve any and all cases in roof framing, is the stepping off method of obtaining the lengths of rafter patterns. The square shown by dotted lines in Fig. 30 gives the position of the square for the first step, as between 1 and 2; the second step would be between 2 and 3; the third step between 3 and 4,

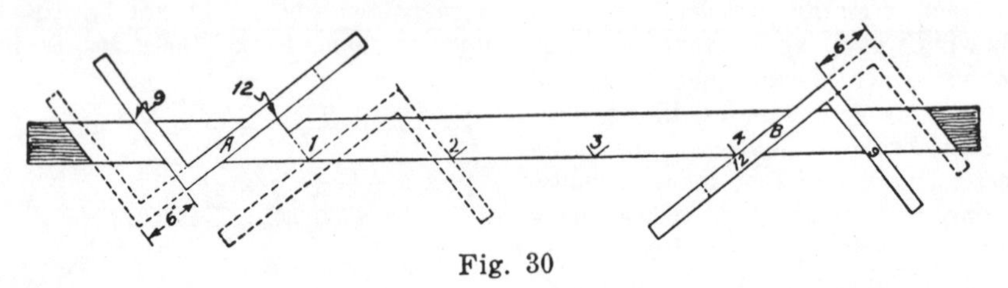

Fig. 30

and the fourth step is shown by the position of the square marked *B*. The figures to be used on the square, to make the pitch conform with the diagram shown in Fig. 28, are 12 on the body of the square and 9 on the tongue, as shown. The diagonal distance between 12 and 9 on the square

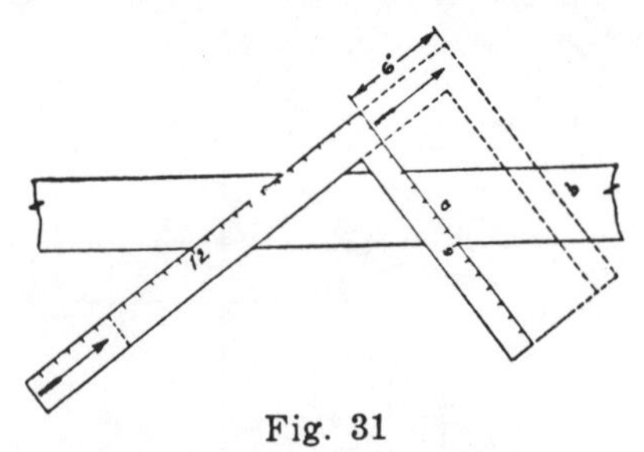

Fig. 31

is 15 inches (see Fig. 29) therefore the rafter would be 15 inches long for every foot run, and the run being 4 feet, the rafter would be 4 times 15 inches, or 60 inches long, which reduced to feet would be 5 feet, the same as we found by the square root method. We are working the problem here to prove its accuracy; in practice, that is not necessary—when the stepping off is done, the problem is solved. But in order to prove that no mistake has been made, the figures used on the square should be checked, and each step should be numbered, as shown, 1, 2, 3 and 4.

Fraction of Step. — The square marked *A*, Fig. 30, is in the position for the first step in stepping off the lookout, or tail of the rafter. If the cornice is to have a 12-inch overhang, then the tail should be marked with the square in the position shown, but if it is to be more, let us say 18 inches, then the square should be moved forward as much as required, in this case 6 inches, as shown by dotted lines, to reach the position for marking the lookout. The shaded part shows what is to be cut off. The same procedure is necessary in case the run is 4 feet 6 inches, instead of just

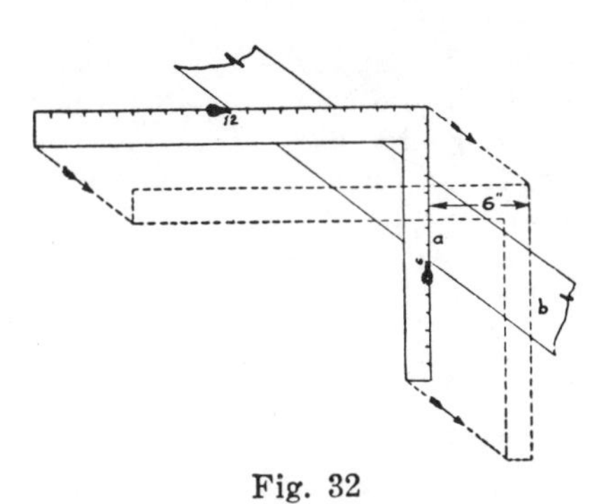

Fig. 32

4 feet. This is shown to the right. The square marked *B* is in the position for the fourth step—to get the additional 6-inch run, move the square forward 6 inches as shown by dotted lines. This is further illustrated in Fig. 31, where, in order to gain the 6 inches in the run, the square is moved forward, as indicated by the arrows, from *a* to *b*. We are using 6 inches for convenience. Any other distance, even fractions of an inch, are handled in the same way.

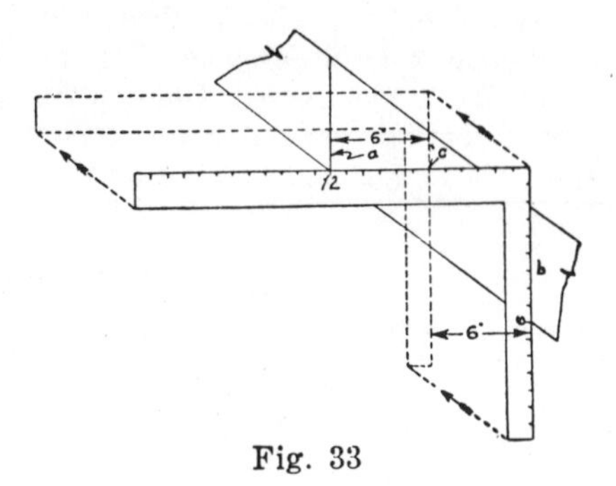

Fig. 33

Other Ways.—How to obtain the same results, when guides are fastened to the square or when a fence is used so that it can not conveniently be moved forward, is shown by Fig. 32. When the last full step is taken, mark along the edge of the tongue of the square, and then slip the square up, as indicated by the arrows and dotted lines, until the square is in the position shown by dotted lines, or from *a* to *b*. The same results can be obtained by making an extra full step, as from *a* to *b*, Fig. 33, and placing a check mark as shown at *c*—then slip the square back, as shown by the dotted lines and arrows, until the edge of the tongue intersects with the check mark. A little study and comparing of these three methods, is necessary in order to understand their relative values.

LESSON 5

STEPPING OFF PROBLEMS AND ROOF CUTS

There isn't anything difficult about the stepping off method for obtaining the length of rafter patterns. And as to the amount of time it takes—that is negligible. Any other method must be studied to be understood so it can be applied without error. Our contention is that the carpenter who solves all of his roof framing problems with

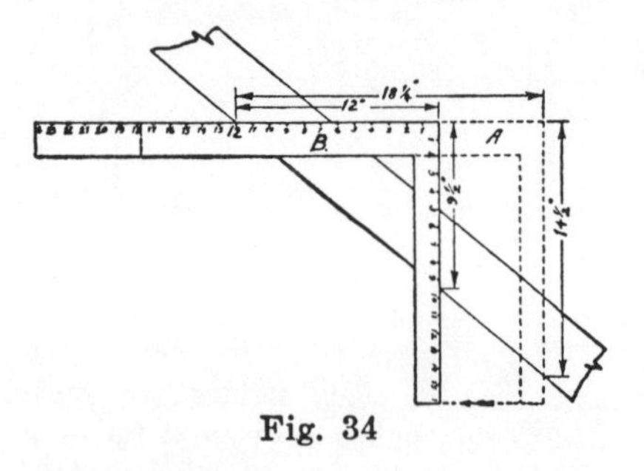

Fig. 34

the steel square, soon becomes so familiar with its applications that he can solve, not only roof framing problems, but any other problems that might come up in carpentry relative to bevels, inclines or angles. Whenever a piece of material must be cut so it will fit in some angular position, the cuts can be obtained with the steel square. Such miscellaneous cuts can never be found in roof framing tables.

Simplifying Problems.—Fig. 34 illustrates a little problem that often must be solved. Suppose we have to install a rafter or a brace with a run of 18 feet 3 inches and a rise of 14 feet 6 inches—how can we simplify the figures to be used on the square in making the cuts and how can the stepping off be done by using the base figure, 12? The square marked *A*, shown by dotted lines, has been applied to a piece of timber (inches representing feet) to obtain the length of the rafter by measuring diagonally between the two figures. But to use 18¼ and 14½ to mark the cuts is rather clumsy. So we mark along the edge of the blade and, keeping the square on this mark, slide it back until the figure 12 intersects with the edge of the timber, as shown by the square marked *B*. The figure on the tongue that intersects with the edge of the timber is the other figure to be used, in this case it is 9½. Then 12 on the blade of the square and 9½ on the tongue will give the cuts for the rafter—the blade giving the foot cut and the tongue the plumb cut. With these figures on the square, the length of the rafter can be obtained by taking 18 full steps, then mark along the blade of the square on the last step and slide the square forward 3 inches and mark the plumb cut along the edge of the tongue. This, if carefully done, is just a little more accurate than the diagonal measurement, with inches representing feet.

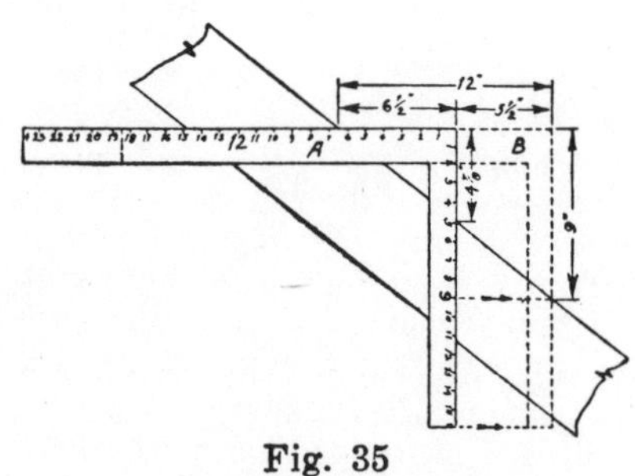

Fig. 35

Changing the Figures on Square.—A problem in reverse order from the one just given is shown by Fig. 35. Here we have a run of 6 feet 6 inches, and a rise of 4 feet 10½ inches. Letting the inches represent feet, we apply the square as shown at *A*, using 6½ on the body and 4⅞ on the tongue.

(Because the figures are small we are using the edge of the timber opposite to what we used in Fig. 34.) Now we mark along the edge of the blade and slide the square forward until the figure 12 intersects with the edge of the timber (in this case 5½ inches) leaving the square in the position marked *B*, shown by dotted lines. This, it will be seen, will give us 9 on the tongue. To obtain the cuts, use 12 on the blade of the square and 9 on the tongue. The blade gives the foot cut and the tongue the plumb cut. The length of the rafter or brace can be obtained by making 6 full steps

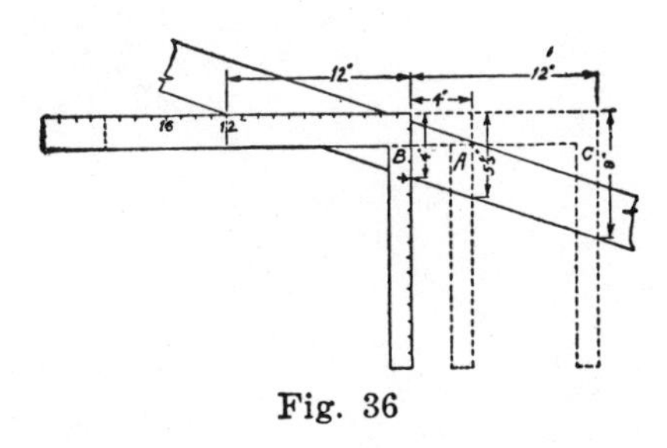

Fig. 36

and sliding the square forward 6 inches beyond the last step and marking the plumb cut along the edge of the tongue.

Low-Pitch Roof Problem. — The problem in Fig. 36 comes up on low-pitch roofs, such as porch roofs. Let us say we have a run of 16 feet and the best we can do with the rise is 5 feet 4 inches. So we apply the square as shown by the dotted lines, marked *A*, mark along the blade, and slide the square back, in this case 4 inches, or until the figure 12 intersects with the edge of the timber. Now the figures will read, 12 on the body and 4 on the tongue. This would require 16 steps to obtain the length of the rafter, however, this can be simplified by slipping the square forward to the position shown at *C*, which would give us 24 on the body of the square and 8 on the tongue. (The same results can be obtained by multiplying both 12 and 4 by 2.) With these figures we would take only 8 double steps in order to obtain the length of the rafter.

The 12-Step Method. — Fig. 37 shows a very simple method of obtaining the length of rafters with

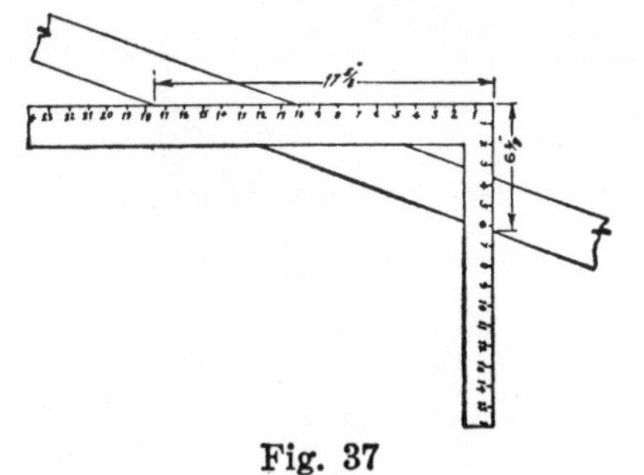

Fig. 37

odd runs and rises—in fact this method works on any kind of run and rise. It is called the 12-step method. For example, we have a run of 17 feet 7½ inches and a rise of 6 feet 4½ inches. Let the inches on the square represent feet, which would give us 17⅝ inches on the body of the square, and 6⅜ inches on the

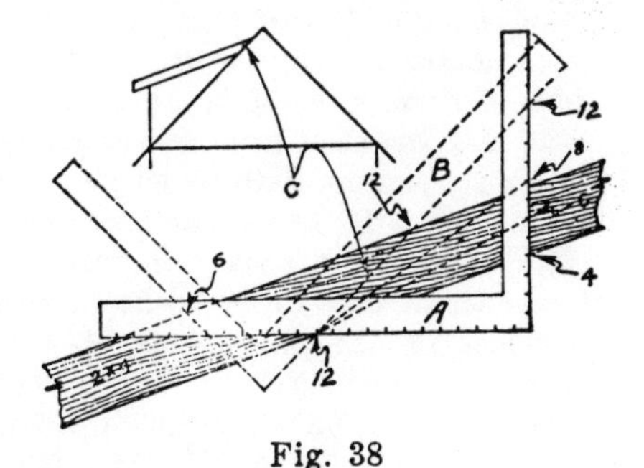

Fig. 38

tongue. With these figures we make 12 steps with the square in order to obtain the length of the rafter. The figures used for stepping off will give the foot and plumb cuts.

The 12-step method just explained is perhaps the simplest method of stepping off rafter patterns, but it is just a little harder to remember than the method where 12 is used as a base figure, or the base-figure method. It should be noticed that in using this method there is no need of sliding the square forward in order to get the fractions of a foot in the run—the fractions are all taken care of in the 12 steps. We are giving this method here as a short-cut method of stepping off rafter patterns, but when the roof involves hips and valleys it is better to stick to the base-figure method.

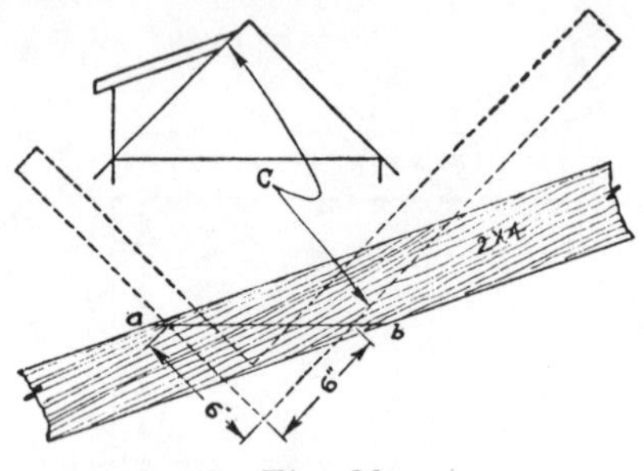

Fig. 39

Low Pitch Joining a Higher Pitch. —Fig. 38 illustrates a problem that often comes up in roof framing, which is to obtain the cut for rafters of a "dutch" dormer that join a main roof —in this instance it is a half-pitch roof. The dormer rafters, let us say, are cut on a 12 and 4 pitch, or one-sixth pitch. The square marked *A* is shown in position to make either the foot or the plumb cuts of the dormer rafter. Now, because the dormer must join a half-pitch roof, the cut would be the diagonal line between 12 on the body of the square and 12 on the tongue, as we are showing by the square marked *B*. It will be noticed that the figures to be used on the square, if applied directly to the timber, would be 12 on the body of the square and 6 on the tongue, as shown by the dotted outline of a square. The arrows, *C*, point out the cut both on the timber and on the diagram. The dotted lines running from the figure 12, square *A*, to figures 6 and 8 show the cuts for the dormer rafter in case it would join a one-fourth or a one-third pitch roof, respectively. The square marked *B* would run from 12 to 6 or 8, instead of, as shown, from 12 to 12. This principle holds good in all pitches.

The same problem is shown solved in a different way by Fig. 39. Here we simply mark the foot or horizontal cut on the timber, as shown from *a* to *b*, and apply the square, using figures that would give the cuts for a half pitch roof, which in this case would be 12 and 12. But because the line isn't long enough to use these figures, we divided them by 2, which gives us 6 and 6, as shown. The cut that would fit a one-half pitch roof is pointed out at *C*, both on the timber and in

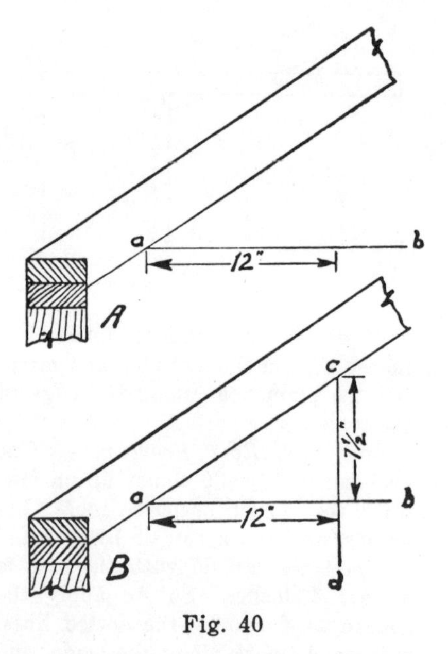

Fig. 40

the diagram to the upper left. Cuts for other pitches are obtained on the same basis.

Obtaining Unknown Cut.—Fig. 40 shows how to proceed to obtain the cut of a rafter in a roof when the pitch is not known. The first operation is shown at *A*, where with a level we obtain the horizontal line, *a-b*. Then we mark off 12 inches, as shown. Now we mark the plumb line, *c-d*, shown at *B*, in such a manner that it will cross the 12-inch point marked off in *A*, which gives us the figures to use on the tongue of the square, or as shown, 7½ inches. So we see that the figures to use on the square are 12 on the body and 7½ on the tongue, the body giving the horizontal cut and the tongue the plumb cut.

LESSON 6

APPLICATION OF SQUARE AND CUTS

Everything in roof framing depends on the common rafter — in other words, the cuts of the common rafter give the basis for all other cuts. If this simple principle is kept in mind, it will be much easier for the student to master roof framing. Another thing to remember is that the carpenter solves his geometrical problems with the steel square and a pencil, rather than with a pencil and paper. And where the student of geometry solves his problems by means of formulas, the carpenter solves the same problems by the application of the steel square. Therefore it is my purpose in these lessons to present various

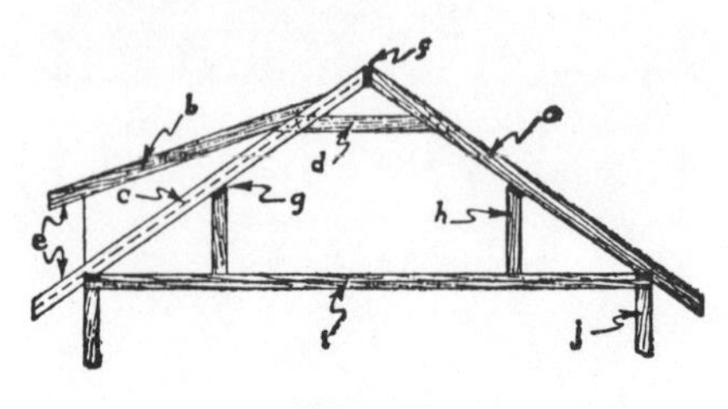

Fig. 41

applications of the square to show how different practical problems can be solved by its use with ease and accuracy.

Names of Roof Members.—Fig. 41 shows a cross section of a gable roof with a "dutch" dormer, ready for the roof sheeting. The pitch of the main roof is a one-third pitch, while the pitch for the dormer is a one-sixth pitch. The names of the different members of this roof are: *a*, common rafter; *b*, dormer rafter; *c*, measuring line; *d*, collar beam; *e*, tails or lookouts; *f*, ridge board; *g*, purlin plate; *h*, purlin studding; *i*, double plate, and *j*, wall studding.

Allowance for Ridge Board.—When a ridge board is used, as shown at *f*, of Fig. 41, one-half of its thickness must be deducted from the run in order to get the proper length of the common rafter. How this can be done is shown by Fig. 42, where the dotted

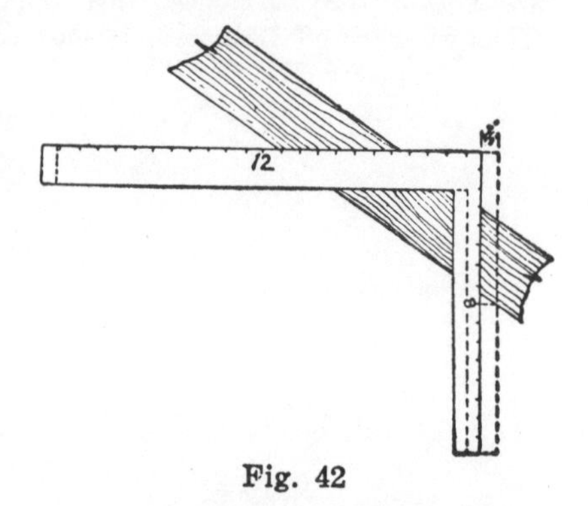

Fig. 42

lines represent the application of the square for the last step, but in order to deduct ⅞-inch from the run, we mark along the blade of the square and move the square back ⅞ of an inch, as shown by figures on the drawing.

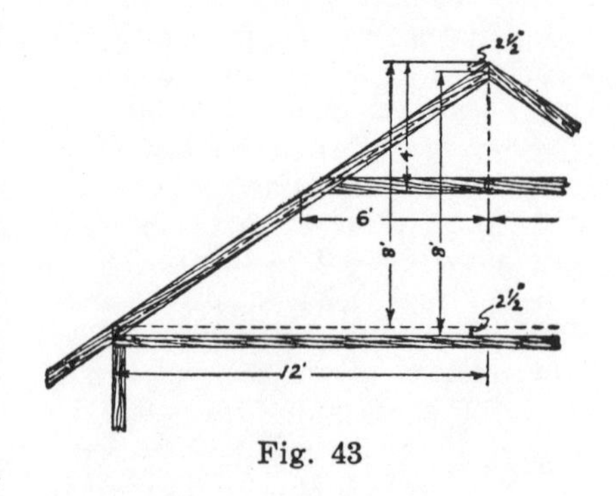

Fig. 43

If one-half of the ridge board is less or more than ⅞ of an inch the deduction should be made accordingly.

Length of Collar Beam and Location.—Fig. 43 shows in part, a pair of rafters in place. The run is 12 feet and the rise is 8 feet. If a measuring line is used the rise begins at the top of the plate and ends 2½ inches below the point of the comb; but if the edge of the rafter is used instead of the measuring line, then

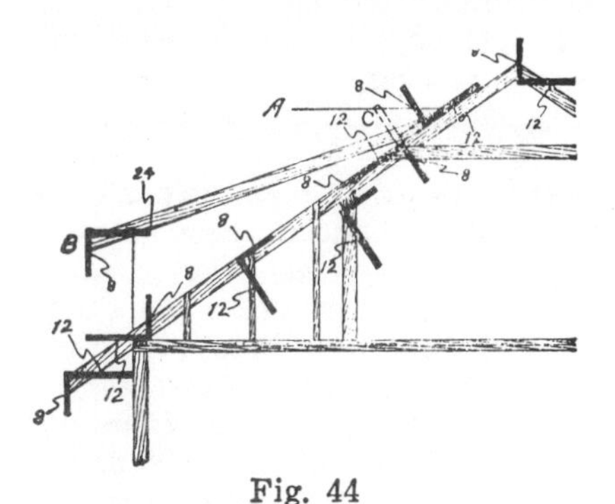

Fig. 44

the rise begins 2½ inches above the top of the plate, as indicated by dotted line, and ends at the point of the comb. This difference often causes confusion in the minds of beginners until it is properly understood. The collar beam is shown just 4 feet below the point of the comb. Having this distance we can determine the length of the collar beam by multiplying 4 by 12, giving us 48, or the distance in inches. This figure is then divided by the rise per foot run, or 8, which gives us the number of feet in the run of a 4-foot rise, or 6 feet. This being a double-pitch roof, we would have to multiply 6 by 2 in order to obtain the full length of the collar beam, which would be 12 feet. By giving this problem a little study it will become clear that when the location of the collar beam is known, its length can easily be determined.

Reading Figures on Square.—Before we take up the next figure I want to discuss the reading of the figures used on a square for framing rafters. The question is, should the figure representing the rise be read first or should the base figure be read first? While I have used these figures both ways, I hold that the base figure should be read first, as 12 and 8 for a one-third pitch; 12 and 6 for a one-fourth pitch, 12 and 24 for a full pitch—the same order should be observed for all other pitches. It is reasonable that the base figure should be read first, for on it rest all the roof framing problems. This reading is supported by the universal standard of reading the width first, and since the run represents the width, it should be read first. It is true that the base figure could be omitted in the reading, or just assumed, but this is not as explicit as reading both figures. There seems to be a justification for reading the rise first when the emphasis is on the rise, as in saying, "the pitch is 8½ to 12," which, if written out in full would be, "8½ inches rise to 12 inches run." If, however, the emphasis is on the cut, then it should read, 12 and 8½, as in this expression: "The cut is obtained by using 12 and 8½ on the square."

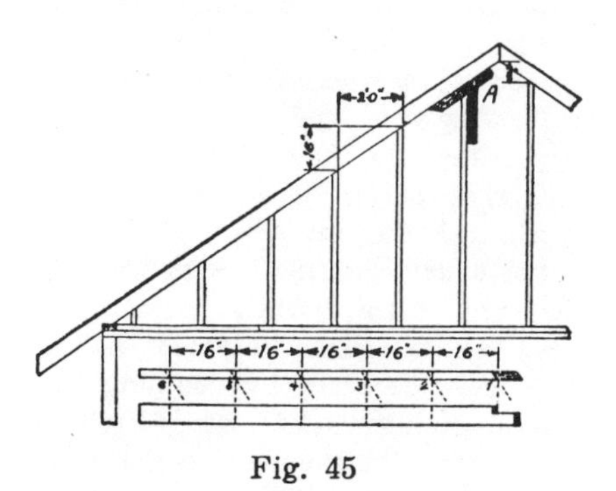

Fig. 45

Application of Square.—In Fig. 44 we are showing the application of the steel square for making the various cuts relative to a one-third pitch roof. It will be discovered that all of the cuts, excepting the plumb cut of the dormer rafter tail, are made by using the figures 12 and 8 on the square. A simple rule for determining whether the blade or the tongue gives the cut is: For all pitches below a one-half pitch, the sharp cuts are marked by the blade of the square and the dull cuts are marked by the tongue, assuming that the run is always taken on the blade of the square. For cuts above a one-half pitch, the rule is

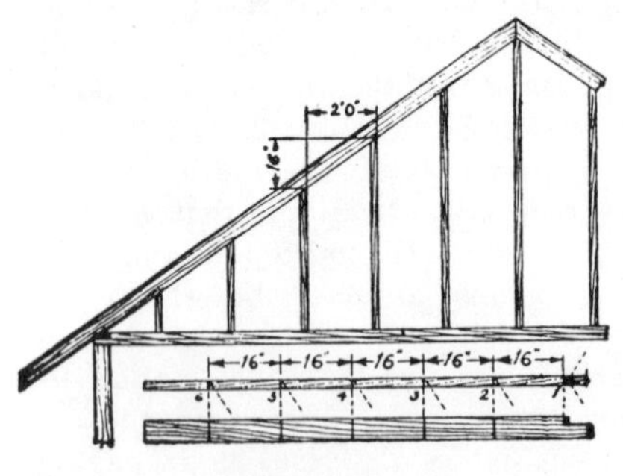

Fig. 46

just the reverse, speaking of cuts relating in some way to the common rafter.

Dormer Rafter Cuts.—The two cuts for the dormer rafter need further explanation. The horizontal line at *A*, simplifies obtaining the cut for the dormer rafter where it joins the main roof. Applying the square to this line, using the figures for a third pitch roof, or 12 and 8, will give the figures for the cut where the edges of the square intersect with the upper edge of the dormer rafter. The same results can be obtained with a little more accuracy by doubling the figures, giving us 24 and 16, which would bring the square into the position shown by dotted lines at *C*. At *B* we are showing how the tail cut for the dormer rafter is obtained. Because this rafter is cut on a 12 and 4 pitch, we have simplified the matter by doubling both figures; hence, we use 24 and 8 to obtain the cut.

Gable Studding.—Fig. 45 shows in part a gable of a one-third pitch roof with the studding in place. The studding are spaced 2 feet on center, which would make the longest pair of studding just 8 inches shorter than the distance between the bottom of the comb joint and the top of the plate, as shown on the drawing to the upper right of *A*. After having determined the length of the longest pair of studding, each of the other pairs would have to be just 16 inches shorter than the pair before. This is illustrated at the bottom, where we show the first pair—the upper one gives an edge view and the bottom one a side view. It will be noticed that these two are cut in pairs. The bevel square at *A*, shows how to set the square for marking the bevels, and the dotted lines at 1, 2, 3, 4, 5 and 6, show how the marking is done.

To the left of the square shown at *A*, we are showing why each pair of studding must be 16 inches shorter than the preceding pair. The studding are spaced 2 feet on center, which gives us an 8-inch drop for each foot, or 16 inches, as shown.

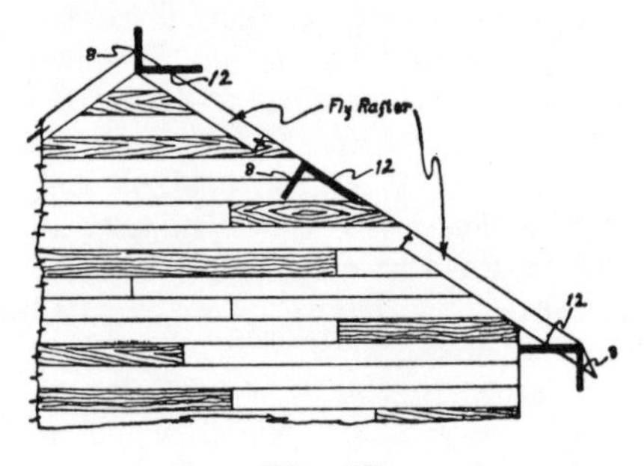

Fig. 47

Another Gable Studding Method.—Fig. 46 shows another method of studding a gable. Here a studding is set at the center, directly under the comb joint. This studding is used for the pattern—two views of it are shown at the bottom. The rest of the studding are cut in pairs, each succeeding pair is cut 16 inches shorter, taking the center studding for the starter. Figs. 45 and 46 should be compared and studied.

Cuts for Gable Boxing and Barge Board.—Fig. 47 shows the application of the steel square for marking the cuts for the boxing and also for the cuts of the fly rafter, or as it is sometimes called, the barge rafter. In all of these cuts the figures 12 and 8 are used, even for obtaining the flare on the end of the tail, as we are showing.

LESSON 7

HIP AND VALLEY RAFTERS

Roof Framing an Achievement.—Occasionally one meets a carpenter who understands the framing of common rafters, but when it comes to hips and valleys he is puzzled. He goes ahead on these rafters more nearly on a trial and error basis, which is to say, that by fitting and trying he manages to get the roof framed, but never knowing just how it was accomplished. The principal reason for this lack of roof-framing knowledge lies with the man himself. He goes about qualifying himself for roof framing like the man who couldn't fix his leaky roof while it was raining, and when it wasn't raining it didn't need it. Roof framing is an achievement—it must be acquired by practice and experience—it is not inherited. So let every apprentice get this thoroughly into his thinking, that while the framing of hips and valleys isn't any harder than the framing of common rafters, it is necessary that one give this part of roof framing more thought and study than it takes for framing the common rafter.

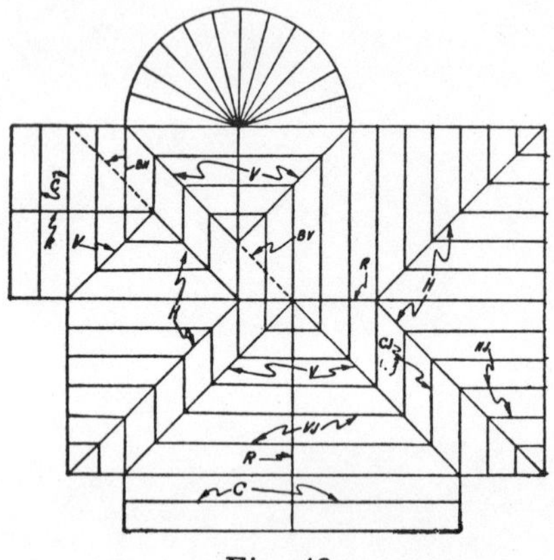

Fig. 48

Names of Rafters.—In order to fix the various kinds of rafters in the reader's mind, we are giving a one-line drawing of a rather complex roof in Fig. 48. This drawing is in plan and does not give the pitch of the roof. We are pointing out with reference letters and indicators the different kinds of rafters in this roof. *C* indicates common rafters; *H*, hip rafters; *V*, valley rafters; *CJ*, cripple jacks; *VJ*, valley jacks; *HJ*, hip jacks; *BH*, blind hip; *BV*, blind valley, and *R*, ridge. The rafters for the circular part can either be framed as common rafters or as jacks, depending on the center support.

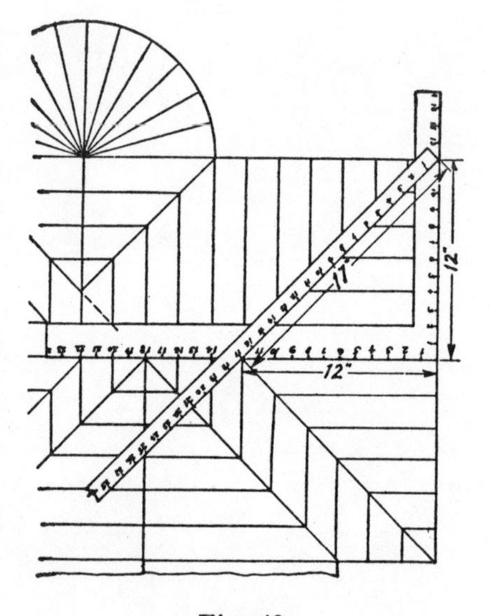

Fig. 49

Run of Hips and Valleys.—The part to the right of the roof we have just been dealing with is reproduced in Fig. 49, where we are showing the difference in the run of the common rafter and the hip or valley. The run of this part of the roof is 12 feet, so by applying the square as shown, letting inches represent feet, we find that the run for the hip rafter is 17 inches, which if brought to full scale would be 17 feet. Because this is true, the reader must remember that when he uses 12 as a base figure in framing

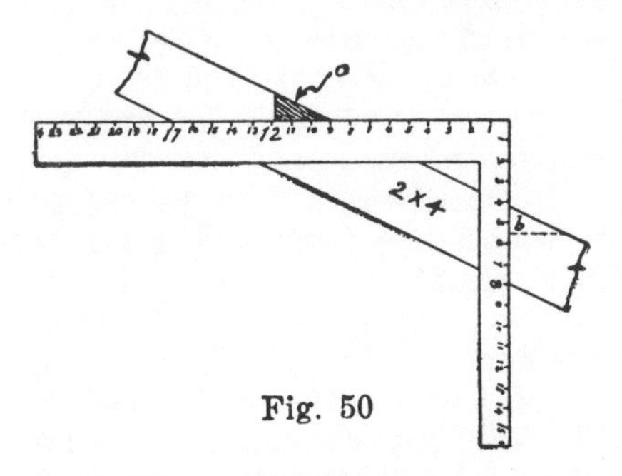

Fig. 50

the common rafter, in framing the hip or the valley he must use 17 as the base, assuming the rafter runs on a 45-degree angle.

Diagonal of 12 and 12.—At this point it should be explained that the diagonal distance between 12 and 12 on the square is not exactly 17 inches, but nearly enough so for most practical purposes. The exact distance is 16.9705 inches, or just a trifle less than 1/32 of an inch less than 17 inches. This difference is not noticeable in marking the cuts for hip and valley rafters, but in stepping off, the run will be increased almost 1/32 of an inch for each step; therefore, if we have a rafter requiring 32 steps we would have increased the run nearly 1 inch, and the rafter would be more than 1 inch longer than the actual requirements. This difference will account for the fact that hips and valleys usually run a little long, when they are stepped off to obtain the length. Some of these variations, though, are due to the inevitable slight variations in the squareness of the buildings or in their not being perfectly level. And since it is almost impossible to escape these variations entirely, it can be said that the extra length gained by stepping off hips and valleys is really a "good fault," for it always leaves something to cut on in making the joint fit.

Seat Cut and Stepping Off Hip and Valley Rafters.—Fig. 50 shows the steel square applied to a 2x4 for a one-third pitch, using the figures 17 and 8, indicating that we are working on either a hip or a valley rafter. The shaded part pointed out at *a*, represents the seat and shows how the blade of the square gives the horizontal cut, while at *b* (an imaginary seat) we show how the tongue gives the plumb cut. Excepting that 17 instead of 12 is used on the blade of the square, stepping off a hip or a valley

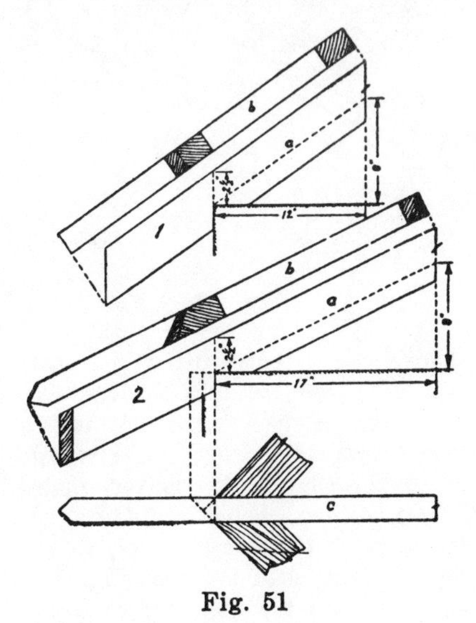

Fig. 51

rafter is the same as stepping off a common rafter, which means that if you take twelve steps for the common rafter, you will also take twelve steps for the hip or for the valley rafter.

Seat Cut of Common and Hip Rafters Compared.—In Fig. 51 we are comparing the seat cut and tail of a common rafter for a one-third pitch roof with the seat and tail of a hip rafter. The common rafter is numbered 1 and the hip rafter, 2. In the five examples shown in Figs. 51 and 52, the *a's* refer to the side views and the *b's* to the bottom views. The

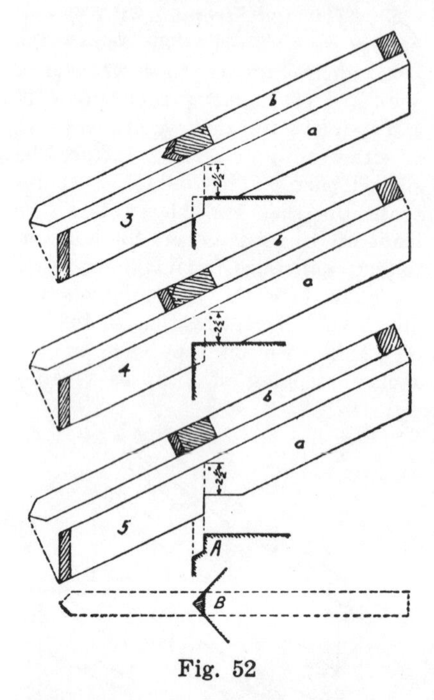

Fig. 52

shaded parts representing the seats should be compared and studied. The difference in the location of the 2½ inches from the corner of the seat to the upper edge of the rafter, shown by dotted lines, should be studied. The two views of the seat of the hip rafter, numbered 2, should be considered in keeping with the plan shown below, marked *c*. It will be noticed that one side of this cut fits tight against the plate, while the other is wide open.

Seat Cut of Hip Rafters.—The seat cut of the hip rafter, numbered 3, Fig. 52, should be compared with the one just considered in Fig. 51. This cut is suitable where the rafter is exposed and a tight joint is required. The seat cut shown in number 4 is open on both sides, as we are indicating by dotted lines on both views. This cut weakens the rafter at the seat more than the cut shown in either number 3 or 5. The seat cut shown in number 5 is perhaps the best, so far as strength is concerned, of those given here. At *A* is shown how the corner of the plate is cut out to receive the rafter. A plan of the plate is shown at *B*, where the hip rafter is represented by dotted lines. The five seat cuts shown in Figs. 51 and 52 should be studied and compared. How to obtain the bevels for these cuts will be taken up in another part of this work.

Seat Cut of Valley Rafters.—Fig. 53 shows two seat cuts for a valley rafter. At *a* is shown the side view—the bottom view of which is shown at *b*. This construction is suitable where a tight fit is required. The bottom view of the seat shown at *c*, leaves a V-shaped opening at the bottom, as

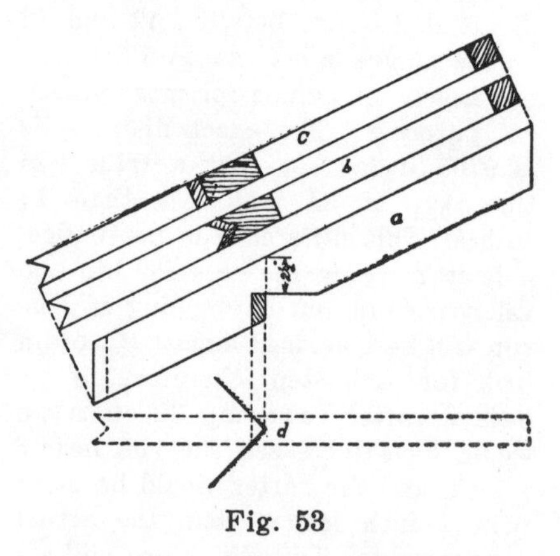

Fig. 53

indicated by the dotted lines. Compare these three views with the plan shown at *d*. The dotted line cutting off a small corner of the angle, shows the opening left if the cut shown at *c* is used for the seat.

Basic Center Line.—It is a foregone

conclusion in roof framing that the framing is technically done on the center line, but in fact the framing is done on the side of the rafter in keeping with the center line. In the common rafter the side of the rafter is in perfect harmony with the center of the rafter, but this is not true in either the hip or the valley rafter. To make this true the hip rafter would have to be backed and the valley rafter would have to have a channel or groove worked into it on the upper edge. This, however, is seldom necessary and rarely done. To be technically correct, both the hip and the valley rafters in reverse order would have to be backed on one edge and grooved on the other in keeping with the pitch of the roof. In rare instances this is a requirement, perhaps, not so much on roofs as on hopper work, which is basically the same as hip roof framing.

LESSON 8

BACKING HIP AND VALLEY RAFTERS

Unbacked Valley Rafters. — Ordinarily in light timber framing, backing of hip rafters or valley rafters is hardly necessary; however, it must be understood in order to frame the seats of those rafters. It must also be taken into consideration when the jack rafters are put in place. An unbacked hip rafter causes no problem when the hip jacks are placed, for if they are kept flush with the top of the hip they are in the right position.

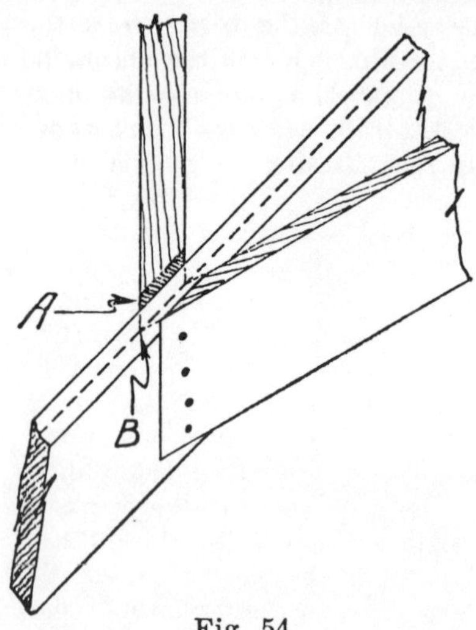

Fig. 54

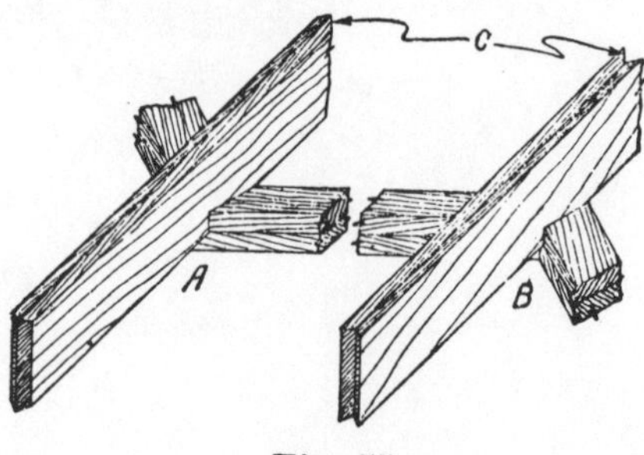

Fig. 55

But this is not true when the backing is omitted on valley rafters. If the valley jacks are fastened flush with the top of the valley it will cause trouble. This is prevented by holding the jacks up above the edge of the valley rafter enough to make the center of the valley come in perfect alignment with the top of the jack and common rafters. At *A*, Fig. 54, we are showing how the jack is held up, and at *B*, by dotted lines, we show how the tops of the jacks line with the center of the valley rafter.

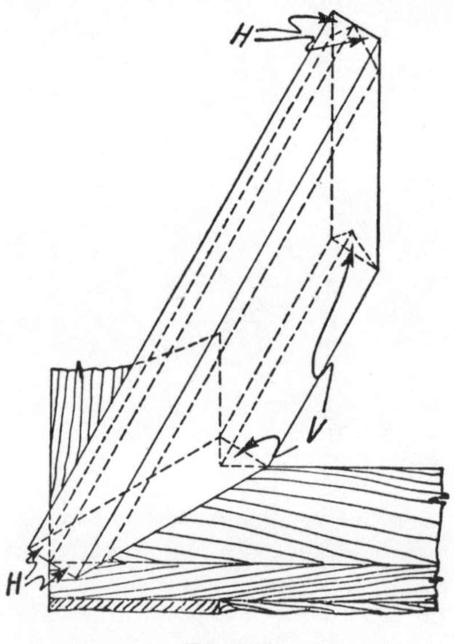

Fig. 56

Joining Jacks to Valley Rafters.— There are a number of practical ways to determine how much the valley jacks are to be kept above the edge of the valley rafter in order to line with both the common rafter and the center of the valley. The experienced roof framer can, in most cases, make the adjustment by eye, but the apprentice should use some means of measuring, either a gauge block or a rule to get the adjustment uniformly the same. Holding a rule on edge on the top of the valley jack and letting it project beyond the cut enough to contact the center of the valley is often used. When the edge of the rule strikes the center of the valley, and at the same time is tight against the jack, the adjustment is correct and the joint can be nailed.

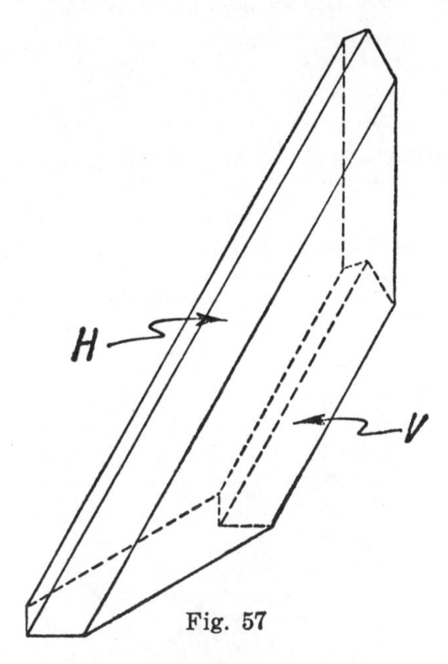

Fig. 57

Backed Hip and Valley Rafters.—Fig. 55 shows a sort of perspective view of, *A*, a backed hip rafter in part, and *B*, a backed valley rafter, also in part. The arrows at *C* point out where the backing of the hip and the backing of the valley are most in contrast, so far as the drawings are concerned.

Practical Way to Back Hips and Valleys.—Fig. 56 shows a practical way of obtaining the depth of the backing for both hips and valleys. Take a block of the rafter material, cut it on the horizontal line of the seat cut and place it on the corner of the plate on a 45-degree angle, as shown, and mark the bottom where it intersects with the plate, which is shown by dotted lines at *H* and at *V* on the foot cut. These lines will give the exact amount of backing required. The dotted lines running from the foot cut to the plumb cut should be studied. The *H*'s point out the backing for the hip, while the *V*'s point out the channel backing for the valley, on the bottom edge of the hip. Fig. 57 shows the block backed for the hip and for the valley.

Method for Backing.—A method of obtaining the backing for hip rafters and for valley rafters is shown by Fig. 58. Here the horizontal cut of the seat is marked near the upper edge of a piece of rafter material, as shown at *a-b*. Then set the points of the compass at half the thickness of the rafter material—with *a* as the center, strike the circle shown by dotted line, marked *X*. This will cut the line *a-b* at *c*. Having this point, mark the depth of the backing as shown between *d* and *e* by dotted line. At *H* is shown an end view of the rafter material backed for the hip, while at *V* is shown the channel backing for the valley.

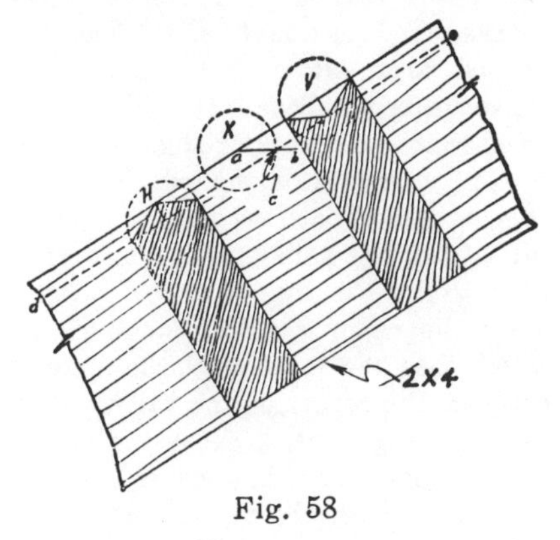

Fig. 58

Marking for Backing with the Square.—A method of obtaining the depth of the backing for hip and valley rafters with the steel square is shown by Fig. 59, in fact, two methods are shown. Take the rise per foot run of the common rafter on the tongue of the square, which in this case is 8, and the length of the hip rafter per foot run of the common

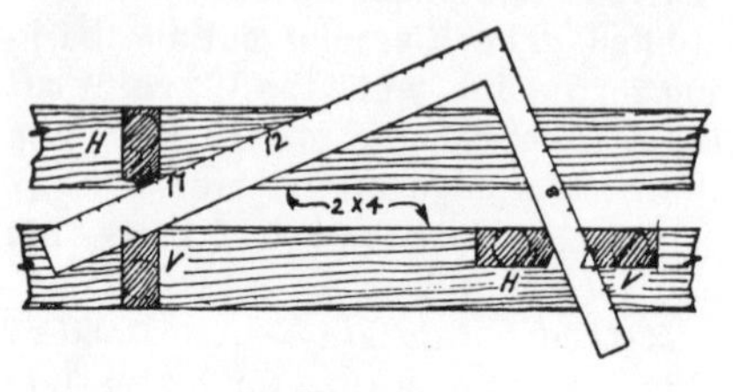

Fig. 59

rafter on the body of the square, or 18¾ inches, and apply the square as we are showing by the upper shaded part, which represents a piece of the rafter material. Both the tongue and the body of the square give the bevel of the backing for both the hip and the valley rafter. The shaded parts marked *H* represent end views of backed hips, while those marked *V* show backed valleys. It will be noticed that one corner of the end views of the valley is unshaded. Without this unshaded part we have the backing of a blind valley, which also represents the backing of valley rafters when the valley rafter is doubled, in which case the joint at the low point becomes the center of the valley. This is frequently done, because the valley rafter must carry the load of the valley jacks. With the hip rafter this is not true, for the hip rafter really carries nothing, because the hip jacks carry the hip and the load of the hip. So there is no necessity at all for increasing the strength of the hip rafter, in fact, it can be made of common rafter material, or even lighter stuff. Study this illustration carefully.

Another Method for Backing.—Still another method of finding bevel and depth of hip and valley rafter backing is illustrated by the diagram shown in Fig. 60, where we are showing a corner of a building. The triangle, *A-B-C,* represents respectively, hip rafter, the run of the hip rafter, and the rise. It will help to clarify the diagram, if the student will imagine this triangle of the hip rafter as lying on its side. With this in mind, at any convenient point strike, at right angle to *A-B, a-b.* Then at right angle to *B-C,* strike *e-d.* Set the compass to

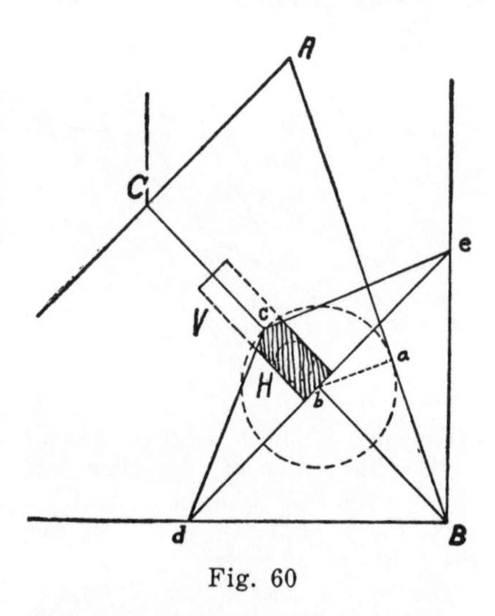

Fig. 60

a-b and strike the circle shown by dotted line, establishing point *c.* Having point *c,* strike both *c-e* and *c-d,* which gives you the bevels of both the hip and the valley backing. Now lay off an end view of both the hip rafter shown at *H,* and the valley rafter shown at *V.* The bevels and the depth of the backing are shown at *c* for both hip and valley rafters.

LESSON 9

STEPPING OFF HIP AND VALLEY RAFTERS

Difference between Hip and Valley Rafters and Common Rafters.—There is only one fundamental difference in the framing of hip and valley rafters and the framing of common rafters, and that is the figures used on the body of the square. In stepping off a hip or a valley you take exactly as many steps as you take in stepping off the common rafter, but instead of using 12 on the body of the square, you use the diagonal distance of 12 and 12, or 17. The figure to be used on the tongue is the same as for the common rafter, which is to say, that if you have a rise of 8 inches per foot run of the common rafter, you also have a rise of 8 inches per the diagonal run of the hip or valley rafters, which is 17. To make this clear in the mind of the student he needs only to imagine the relationship of the run of the common rafter to the run of the hip or valley rafters. We suggest the use of the student's imagination here, because much depends on his ability to imagine, as to whether or not he understands clearly the different problems that invariably will come up in roof framing. Roof framing is primarily a product of the mind—mere muscles and a willingness to use those muscles will not produce accuracy in roof framing. The whole mental process must be utilized in order to frame a roof understandingly.

Stepping Off Hip and Valley Rafters.—Fig. 61 shows how to step off a hip or valley rafter after the seat cut has been made, which was explained in lesson 7. The figures used on the square are 17 and 8, which indicate a one-third pitch roof. The shaded square shown at number 1 is in position for the first step; number 2, the second step and so on until we reach the 5th, or the last step, which is represented by the shaded square to the extreme right. The tongue of the square gives the plumb cut. To the left we are showing how to step off the tail of the hip rafter. If the cornice projects over the building only 12 inches, the tongue of the square in the position shown will give the plumb cut. But if, for instance, the cornice projects 18 inches, then the square must be slipped forward 8½ inches, as shown by the dotted lines, in order to gain the extra 6 inches in the width of the cornice. If the run of the common rafter were 5 feet 6 inches, instead of only 5 feet, the ex-

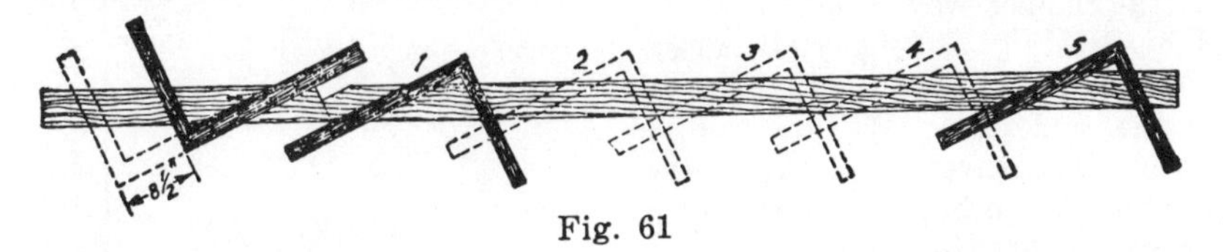

Fig. 61

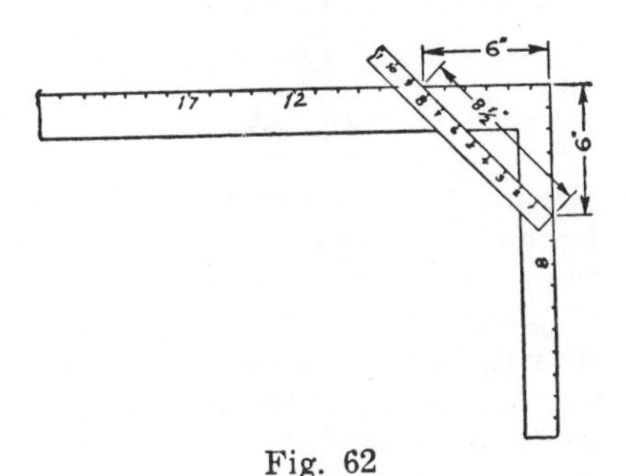

Fig. 62

tra 6 inches of the run would be obtained on the hip rafter by slipping the square forward 8½ inches on the fifth step, just as we have done in gaining the extra 6 inches for the cornice.

Run for Fraction of Step.—Why we used 8½ inches to gain the extra 6 inches in the run of the common rafter is illustrated by Fig. 62. Here it will be seen that the diagonal distance of 6 and 6, on the square, is 8½ inches. A still simpler way to arrive at this is by dividing 17 by two, because 6 is one-half of 12, and the results will be the same, 8½. We are using 6 inches for convenience. Any other number of inches or fractions of inches are treated in the same way. It should be remembered here, that

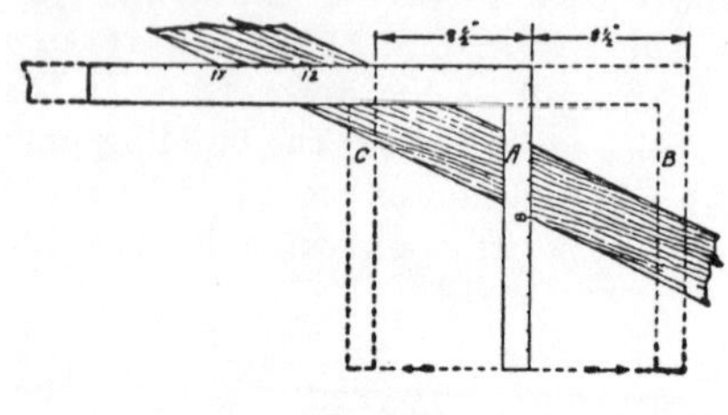

Fig. 63

dividing 17 by the denominator of the fraction of a step, is simple only so long as the fraction is simple.

Taking Fraction of Step.—A detail showing how to take a fraction of a step in stepping off hip or valley rafters is shown by Fig. 63. The square marked *A* is in the position of the last step, and to gain the fraction of a step, which in this case is 8½ inches, we slip the square forward 8½ inches to the position shown by dotted lines and marked *B*. The same results can be obtained by taking an extra step and moving the square back 8½ inches to the position shown at *C*, which is 8½ inches to the right of 17, position *A*. In both of these methods the tongue gives the plumb cut. In simple terms, take the diagonal distance of the fraction of a step of the common rafter and add it to the last full step of the hip or valley rafter as explained in Fig. 63, in order to get the full length of the hip and valley rafters.

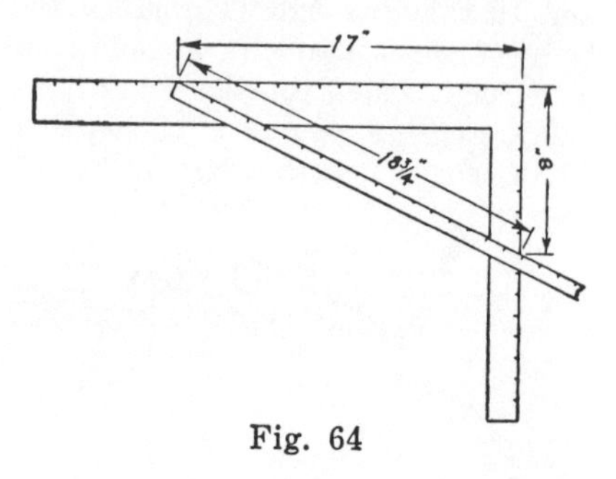

Fig. 64

Edge Bevel for Hip and Valley Rafters.—To obtain the edge bevel for a hip or a valley rafter to fit against a deck or into a right angle, take 17 on the body of the square and the length of the hip rafter per 17 inches of run on the tongue. The latter gives the bevel. I show how to obtain the length of a hip or a valley rafter for a one-third pitch roof in Fig. 64, by

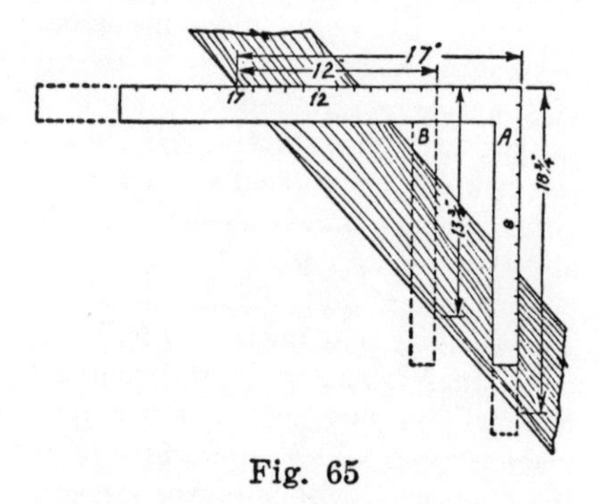

Fig. 65

measuring the diagonal distance of 17 and 8, which gives us 18¾ inches, therefore, 17 and 18¾ (the latter gives the bevel) will give the edge bevel for hips and valleys of a one-third pitch roof.

Reducing Figures on Square—The

reader has no doubt discovered that 18¾ can not be taken on the tongue of the square, in which case the figures should be reduced. Fig. 65 shows one method of reducing the figures. Apply the square as shown at *A* and

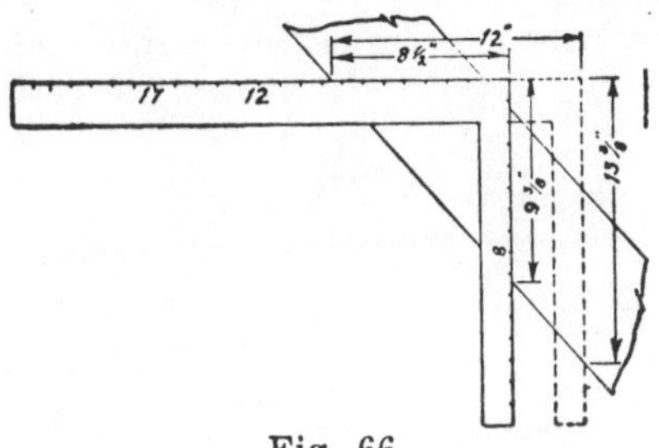

Fig. 66

mark along the blade of the square. This done, pull the square back until the figure 12 intersects with the edge of the timber, or to the position shown by dotted lines at *B*, which will give you the proportional figures, or 12 and 13⅜. These figures will give the edge bevel of the hip or the valley rafter for a one-third pitch roof, the larger figure giving the cut.

Another Way.—The same results can be obtained by dividing 17 and 18¾ by 2, which will give us 8½ and 9⅜. Using these figures, apply the square as shown by Fig. 66—the tongue gives the cut. Now, if the square is pushed forward until the figure 12 intersects with the edge of the timber, as shown by dotted lines, we will again have 12 on the body of the square and 13⅜ on the tongue—the latter gives the cut.

Whys and Wherefores.—We are showing, by Fig. 67, why the run of the hip and the length of the hip rafter per foot run of the common rafter will give the edge bevel of the hip or valley rafters. The drawing represents a plan of one end of a hip roof having a 12-foot run. The square marked *A*, shows how 17 and 17 would give the bevel if the roof had no pitch at all—the square would have to be placed in the position shown by dotted lines at *B* in order to mark the bevel. But our roof has an 8-inch rise per foot run, which would increase the figures on the blade of the square marked *A* to point *a*, or to 18¾. The triangle shown represents: *b-c*, a hip rafter; *c-17*, the run; and *17-b*, the rise. This triangle is shown as if it were lying on the side. The dotted part-circle from *b* to *a* shows how the length of the rafter has been transferred to the blade of the square. Now, if the student will imagine the blade of the square, pivoted at the heel, being lifted at the end until point *a* is directly above point 17, he will see why 17 and the length of

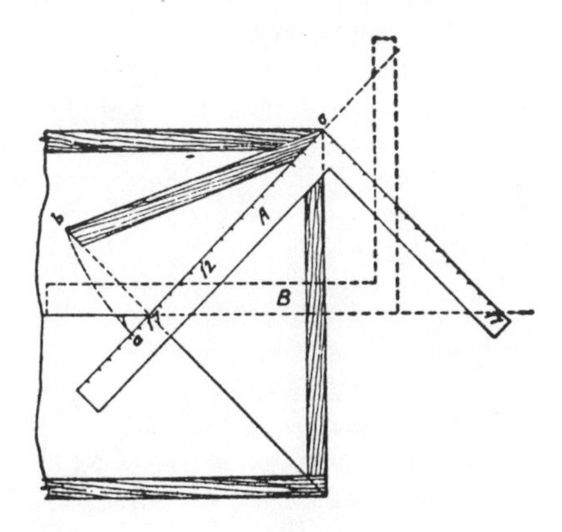

Fig. 67

the rafter will give the edge bevel for the hip and the valley rafters. To mark the cut, the square would have to be placed in the position shown at *B*. To make it convenient, the figures would have to be reduced as we have shown by Figs. 65 and 66.

LESSON 10

DIFFERENCE IN LENGTHS OF JACKS

Principles of Roof Framing.—Almost all of the problems we have been dealing with in these lessons on roof framing have been based on the one-third pitch. This pitch is no more conducive to clarify roof framing problems than any of the other

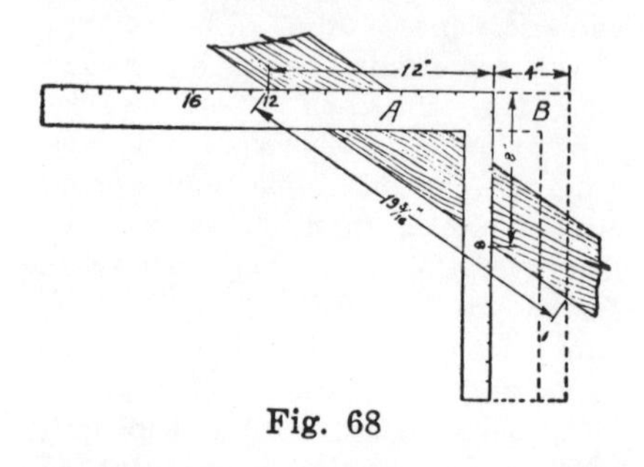

Fig. 68

pitches. We are holding to this pitch so as to simplify the illustrations. After one understands the principles of roof framing as applied to one pitch, he understands the principles that will apply to all pitches. The only difference that will be found is in the rise, for where the one-third

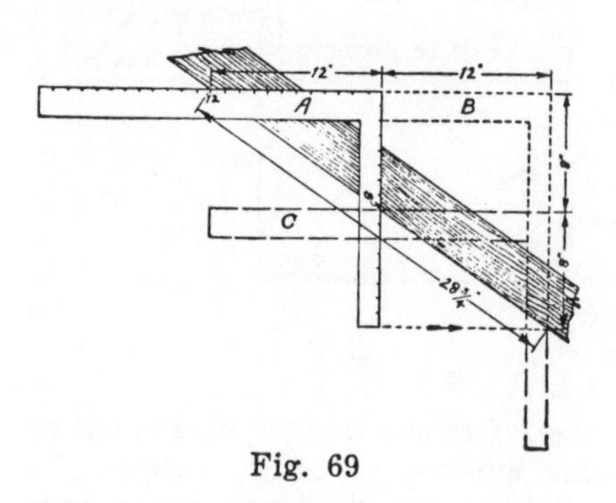

Fig. 69

pitch has 8 inches rise to the foot run of the common rafter, the other pitches can have rises per foot run that go higher or lower than 8 inches, and even into fractions of an inch. In fact, there is no limit to the number of pitches that can be used in roof framing—whenever a roof has a change in the slope, there is a change in the pitch of that roof. As we explained in another part of this work, there are a number of basic

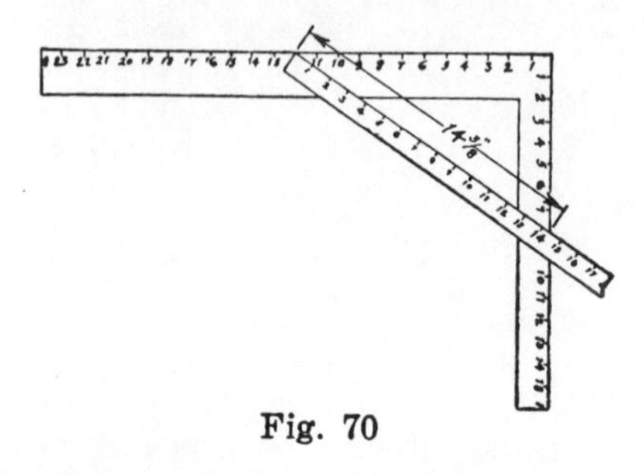

Fig. 70

pitches, as one-third, one-fourth, one-half and so forth, and while these have established a sort of standard, they do not carry with them any special qualities. They are used more frequently than other pitches for convenience rather than because of special merits.

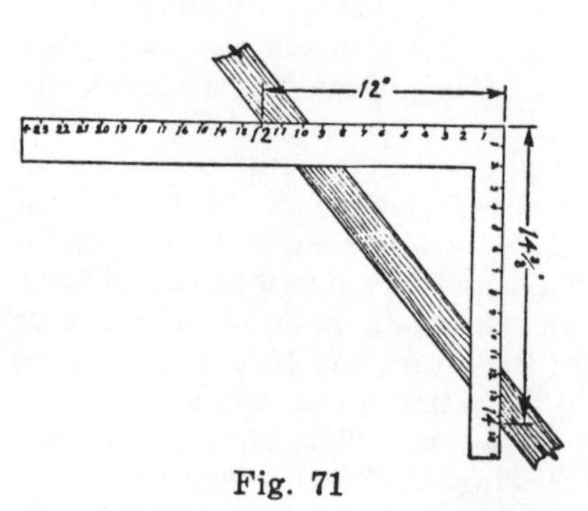

Fig. 71

Jack Rafters.—Jack rafters in reality are common rafters cut to fit against hip or valley rafters. The horizontal and the plumb cuts are the same as for the common rafter. The edge bevel of the side cut and the fact that there is a difference in the lengths of jack rafters, makes them different from the common rafter.

Difference in Lengths of Jacks.—Fig. 68 shows how to determine the difference in the length of jack rafters. The square in the position shown at *A* represents the square applied to a common rafter for one-foot run, and if the rafters were spaced 12 inches on center, the diagonal distance between 12 and 8 would be the difference in the lengths of the jack rafter. But rafters are seldom spaced less than 16 inches on center; therefore, to obtain the difference in the lengths of jack rafters for a 16-inch space, we would move the square forward 4 inches, or to position *B*. Now, as can be seen by the drawing, the difference in the lengths is 19⅜₁₆ inches.

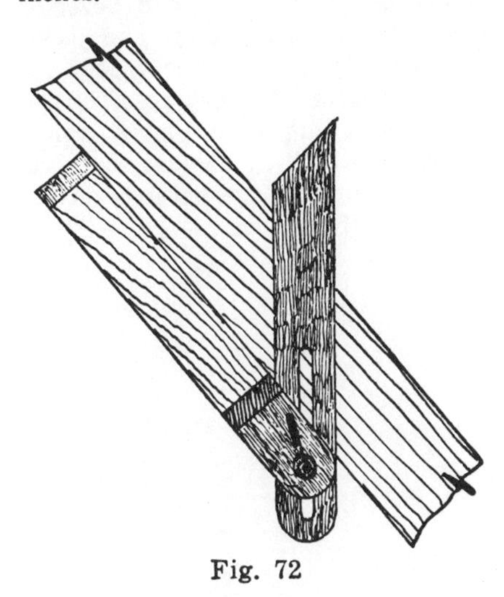

Fig. 72

Jacks Spaced 2 Feet.—How to obtain the difference in the lengths of the jack rafters when the rafters are spaced 2 feet on center is shown by Fig. 69. Here we show two methods. First, the square can be moved forward to position *B*, shown by dotted lines, which will give us the difference in the lengths of the jack rafters, or 28¾ inches, plus. (The exact figures are 28.84.) Second, we can take another step as shown by the square represented by the dashed lines, and the two steps will give us the difference in the lengths of jack rafters spaced 2 feet on center, or, as in the other instance, 28¾ inches, plus. The same results can be obtained without

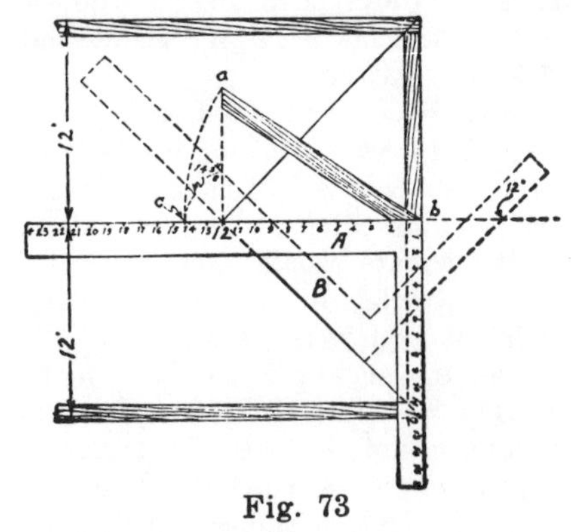

Fig. 73

the square by taking twice the diagonal distance of 12 and 8, which is the same as 14⅜ inches, plus (the exact figures are 14.42) multiplied by 2, or 28¾ inches, plus.

Edge Bevel of Jacks. — Fig. 70 shows how to measure the diagonal distance of 12 and 8 with a rule in order to obtain the length of the common rafter for one foot run, or 14⅜ inches, plus. It is necessary to have this distance when we apply the square to obtain the edge bevel for the side cut of a hip jack or a valley jack. For instance, 12 and the length of the common rafter per foot run (the latter gives the bevel) will give the edge bevel for the side cut of the jack rafters for any pitch. Fig. 71 shows the square applied to the timber for obtaining the edge bevel of jack rafters for a one-third pitch roof. In other words, the edge bevel for the side cut of jack rafters for a one-third pitch is obtained by taking 12 on the body of the square and 14⅜ on the tongue; the tongue gives the bevel.

Marking with Bevel Square.—After the edge bevel has been determined with the square, take a bevel square and set it to the proper angle and use it for marking the side cuts for the jack rafters. Fig. 72 shows a bevel square applied to the timber, giving the edge bevel for a side cut of a jack rafter.

Edge Bevel and Figures Used.—Fig. 73 shows a plan of one end of a hip roof with a 12-foot run. If the roof had no pitch at all, 12 and 12 on the square would give the bevel for the jacks, as we are showing with the square in position *A*. Of course, the square would have to be brought to position *B* before the marking could be done. The triangle: *a-b*, rafter; *b-12*, run, and *12-a*, the rise, represents a rafter lying on its side. The dotted part-circle, *a-c*, shows how the length of the rafter has been transferred to the square, giving us 14⅜ inches, plus. Now if the student will imagine the square marked *A* pivoted at the heel, and the end of the blade lifted until point *c* will be directly over point 12, he will see why 12 and 14⅜ will give the edge bevel for the side cut of jack rafters for a one-third pitch roof. This principle holds good no matter what the pitch of the roof might be. As stated before, the square must be changed to position *B* in order to do the marking.

Side Cuts.—The term, "side cut," or as it is sometimes called, "cheek cut," is often erroneously applied to the edge bevel for making the side cut. The side cut has two bevels, the edge bevel and the side bevel (the plumb cut) but the cut itself is that part, speaking of jack rafters, that fits against a hip or a valley rafter. Sometimes the edge bevel is referred to as "top cut," which again is erroneous, for the edge bevel of side cuts for hips, valleys and jacks, do not always come on the upper edge of the timber. In making a bird's-mouth seat cut on a hip or valley rafter so it will fit against the plate or a purlin, the edge bevel is on the bottom of the timber.

LESSON 11

IRREGULAR-PLAN ROOFS

Less Frequently Used.—In the previous lessons on roof framing we dealt with problems relating to regular roof framing, which might be called common roof framing, because it is more commonly used than what

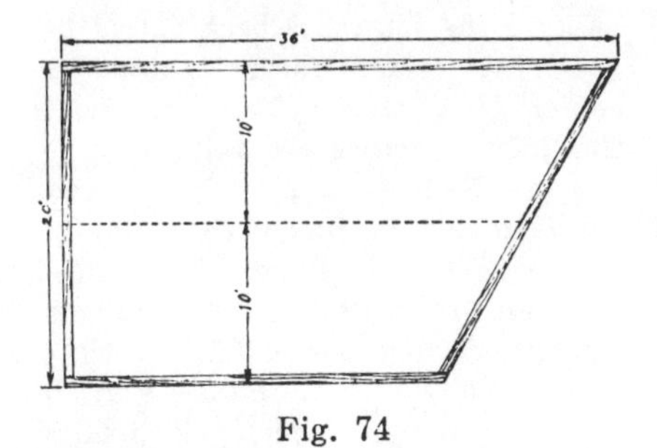

Fig. 74

is known as irregular roof framing. The two branches of roof framing are fundamentally the same; however, the irregular roof, whether it is irregular in plan or in pitch, is less frequently used. For that reason the framing of such roofs apparently is more difficult than the framing of a regular roof.

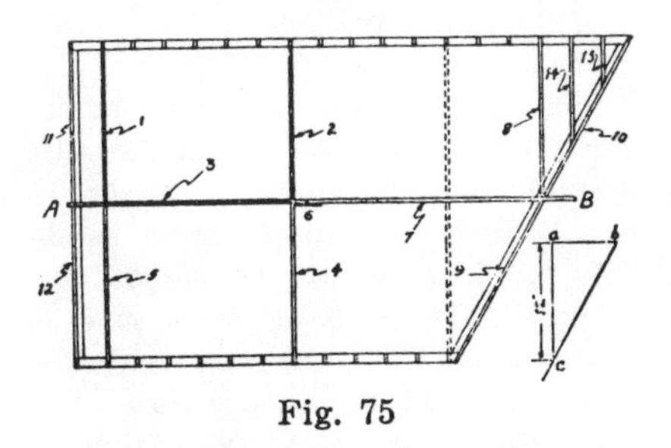

Fig. 75

Simple Problems First.—In dealing with the problems that must be solved in irregular roof framing we intend to be as practical as possible in order to get the problems across to the readers. As our practice has been, we will take up simple understandable problems first and, step by step, lead up to the more difficult problems. We are assuming that the reader understands what is meant by the terms, "run", "rise", "pitch", "span" and so forth, all of which have been covered in previous lessons. The emphasis from here on must be placed on problems pertaining to the irregular roof, both in plan and in pitch.

Raising Rafters on Irregular-Plan Roof.—Fig. 74 is a simple plan, irregular on one end, which is to receive a regular pitch roof. The dotted line represents the comb of the roof. Fig. 75 shows the plan with the plates spaced for the rafters and some of the rafters in place. The ridgeboard should be spaced in keeping with the plates. In framing this roof we would proceed by nailing the ridgeboard, numbered 3, to the two rafters numbered 1 and 2. Then raise these rafters, holding them in place with rafters number 4 and 5, and brace this part. Then ridgeboard number 7 should be nailed to rafter number 8 and raised, making the ridge joint with a cleat as shown

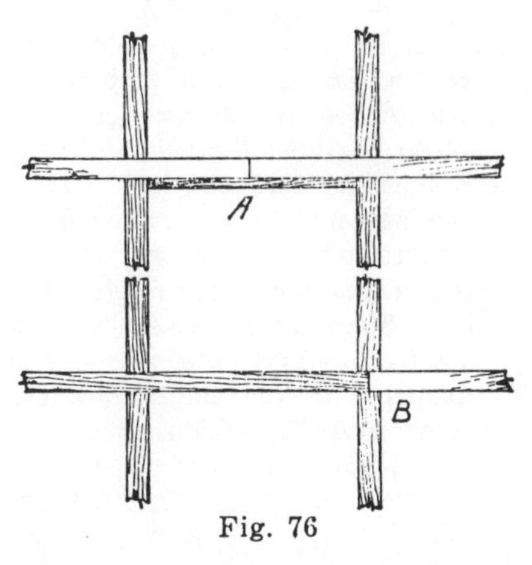

Fig. 76

at 6. Rafter number 9 should then be placed to hold the ridgeboard up. (An equally satisfactory way would be to place the rafters shown by dotted lines before number 8 and number 9 are placed.) Then place number 10 followed by 11, 12, 13,

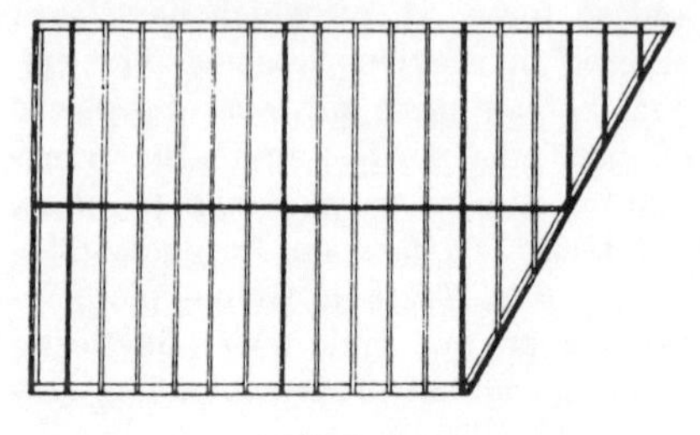

Fig. 77

14 and so on until the rafters are all up. At *A* and *B* we show the ridgeboard projecting beyond the plate lines so that the end rafters can be adjusted before the ends of the ridgeboard are cut off.

Obtaining Edge Bevel of Jacks.—How to obtain the edge bevel for the jack rafters on the irregular end is illustrated with the diagram to the right of Fig. 75, which should be

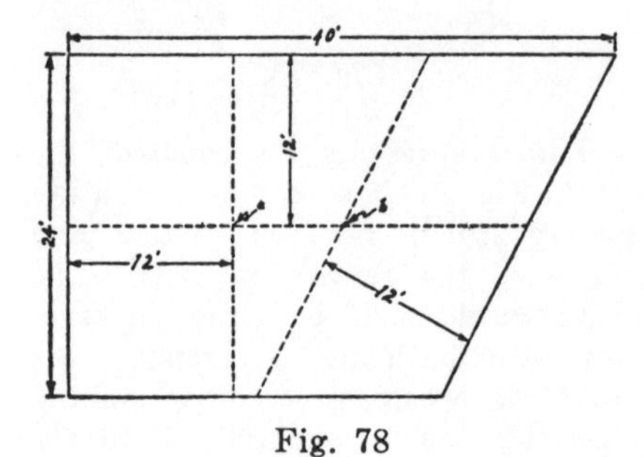

Fig. 78

drawn to full scale. The distance *a-b* and the length of the rafter per the distance *a-c*, will give the edge bevel of the jack rafters—the rafter length giving the cut. (The distance *a-c* represents a foot run.) The edge bevel for the irregular end rafters, where they join the ridge, is obtained as explained in Figs. 106, 107, and 108.

Fastening Ridge Board.—Fig. 76, *A*, shows a detail of the method of fastening the ridge joint shown at 6 in Fig. 75. At *B* is shown the commonly used joint. Either of these joints give satisfactory results, however, the one shown at *A* will leave the nailing for the rafters unimpaired because the joint has been placed between the rafters rather than at the center of a rafter. The cleat holds the joint together securely.

Fig. 77 shows a plan of the roof we have been dealing with having all of the rafters in place. The shaded members represent the preliminary work in raising the roof.

Hips on Irregular Plan.—Fig. 78 shows an irregular plan which is to

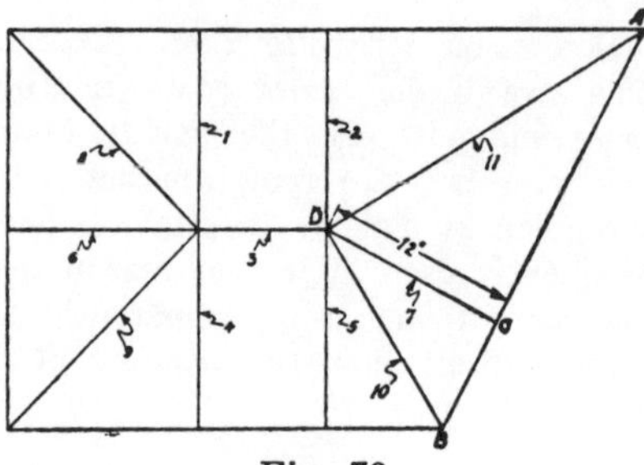

Fig. 79

receive a hip roof. The run, as indicated by the figures is 12 feet. By drawing the dotted lines parallel with and 12 feet from the plates we establish the location of the ridge from point *a* to point *b*. Fig. 79 gives a diagram of the same plan with the principal rafters in place. To raise these rafters we would nail the ridgeboard, numbered 3, to rafters 1 and 2, then raise this and place rafters 4 and 5—then 6, 7, 8, 9, 10 and 11. Now the remaining rafters can be placed. The hips and jacks to the right are irregular, while the rest of the roof is a regular hip roof.

Edge Bevel of Jacks for Long and Short Hips.—Fig. 78 shows that the run of this roof is 12 feet, but in the diagram (Fig. 79) we will let inches represent feet, giving us a run of 12 inches. Now to obtain the edge bevel for the jacks that join the long hip, take the distance from *A* to *C* on

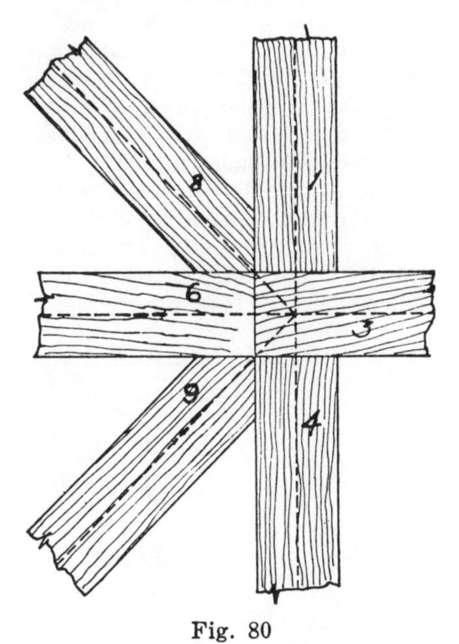

Fig. 80

the body of the square and the length of the common rafter for the distance *C-D*, on the tongue; the tongue gives the cut. (*C-D* represents a foot run.) To obtain the edge bevel for the jacks that join the short hip, proceed by taking the distance *C-B* on the tongue of the square, and the length of the rafter for the distance *C-D* on the body; the body gives the cut. How to obtain the bevels for the hips will be taken up in a later lesson.

Joining Hips at Ridge.—Fig. 80 is a detail giving the joints at the end of the ridge to the left. The dotted lines represent the centers of the various members, and it should always be remembered that roof framing is done on a basis of the center line. These joints represent regular hip roof framing. Fig. 81 shows the joints to the right, which are of a regular-pitch hip roof on an irregular plan. The figures shown in both details are the same as the figures used on the diagram shown in Fig. 79. The two details should be studied and compared.

Blunders.—The men who are fired because of blunders made in simple matters far outnumber those who lose their jobs because of blunders made in difficult tasks. This is logical reasoning—simple things are neglected because they are so simple, and then suddenly one comes up for solution and we blunder, just because it was too simple for consideration. One needs only to observe new workmen going about their tasks to dis-

Fig. 81

cover that more blunders are made in simple matters than in difficult ones. The reason for this marked difference is that when one has a difficult task to perform, he gives the matter consideration, but if it is simple, he neglects it.

LESSON 12

IRREGULAR DECKS AND RAFTER CUTS

Diagram Most Practical.—The irregular roof, whether in plan or in pitch, is rarely used; however, when it becomes necessary to use it, then, as a rule, there is no other solution for the problem. We are using for convenience, as nearly as possible the same plan for illustrating the various problems that we wish to bring before our readers; that is, the dimensions and the plans are kept as nearly alike as we can. In actual practice it is seldom that the same plan is used twice, for the simple reason that the irregular plan is mostly used as a solution to some problem brought about by existing circumstances. What we want to impress on the readers' mind is that if they understand thoroughly the principles involved in framing a roof for a certain irregular plan, they will have no trouble applying those principles to any irregular plan roof that they might be called on to frame. Making a diagram of the roof plan on as large a scale as convenient, we believe, is the most practical way of obtaining the various cuts and lengths of the different rafters of either the irregular-plan roof or the irregular-pitch roof. While the square root method is more accurate, it is not the most practical, since so few carpenters are able to work a problem in square root off-hand. A diagram in roof framing, to be exact, is simply materialized square root.

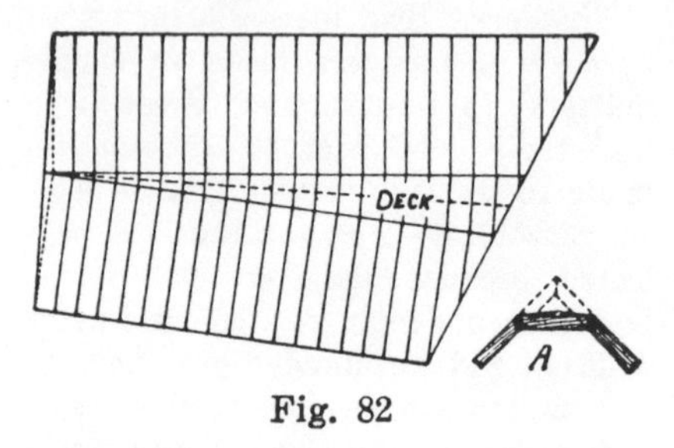

Fig. 82

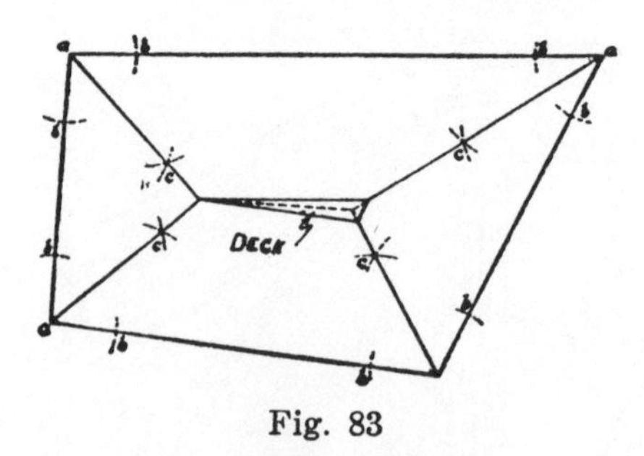

Fig. 83

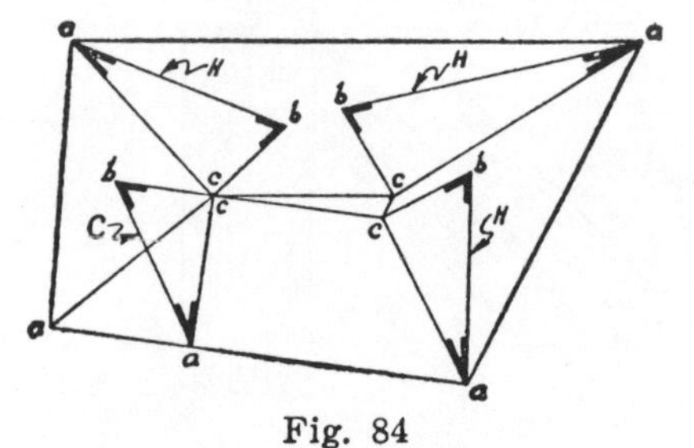

Fig. 84

Irregular Decks.—Fig. 82 is a one-line drawing of a plan of a roof that is irregular in plan throughout. By using a deck, as we are showing on the plan, the framing of the roof becomes a matter of framing common rafters for a regular-pitch roof, excepting that the jack rafters and the end rafters on either end are irregular and must be framed accordingly. The dotted lines shown to the left indicate a pair of regular common rafters. Technically speaking, the end rafters (left) must be cut just a trifle longer with a slight bevel; however, in practice, on this plan, a pair of common rafters would answer the purpose, because the difference in the lengths would hardly be noticeable. If the irregularity in the plan were greater, then the end rafters would have to be framed to the proper length and bevels. The end rafters at the right and the jack rafters would be framed as explained under Fig. 75 of the last lesson. At *A* we show a detail of the deck with a tapering fillet to give the deck fall. The

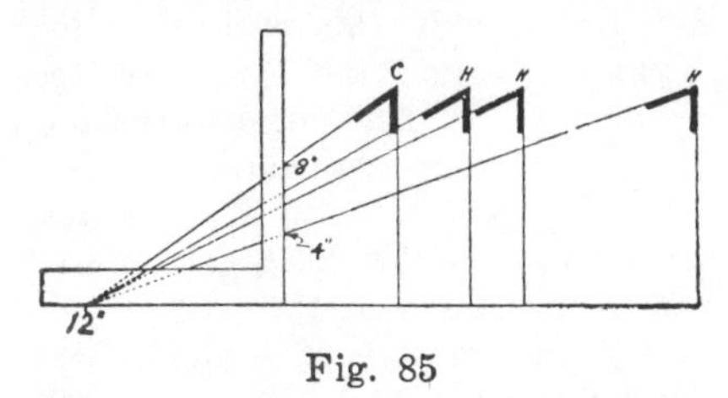

Fig. 85

dotted lines show how the roof could be framed without a deck, in which case each pair of rafters would have to be framed a little shorter than the pair before. To determine the difference in the lengths of the rafters, a pattern for the shortest pair of rafters and a pattern for the longest pair of rafters should be made and the difference in the lengths of these rafters should be divided into as many equal parts as there are

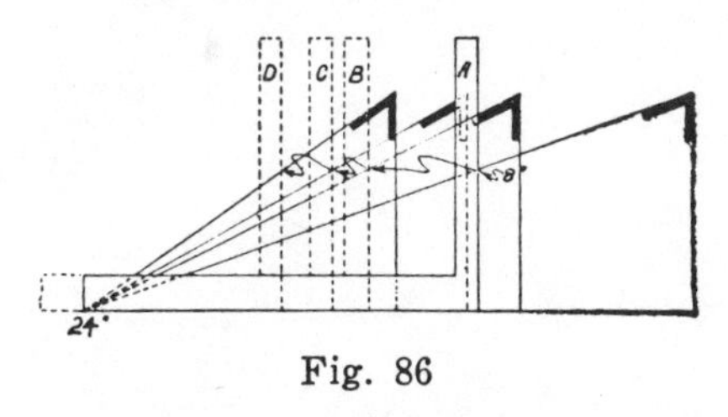

Fig. 86

spaces on the plan between the longest and the shortest rafters, minus one, which will give you the different lengths of all of the rafters. The dotted line through the center shows where the comb would come, which would incline toward the left end.

Irregular-Plan Hip Roof.—Fig. 83 shows the same plan but for a hip roof. If the deck that is shown were omitted the roof would come to a point at the dotted line. The positions of the hips are found by setting one leg of the compass at *a* and, at a convenient distance, striking *b*, *b*. Then strike the cross at *c* from *b*, *b* and draw the hip from *a* through the cross at *c*.

Diagram Showing Rafters.—Fig. 84 is a diagram showing three hip rafters and one common rafter—the *H*'s indicate hips and the *C* a common rafter. In each of these rafters *a-b* represents the rafter; *c-b*, the rise and *a-c*, the run. To get the right conception, imagine each of these rafters in triangular form, hinged at

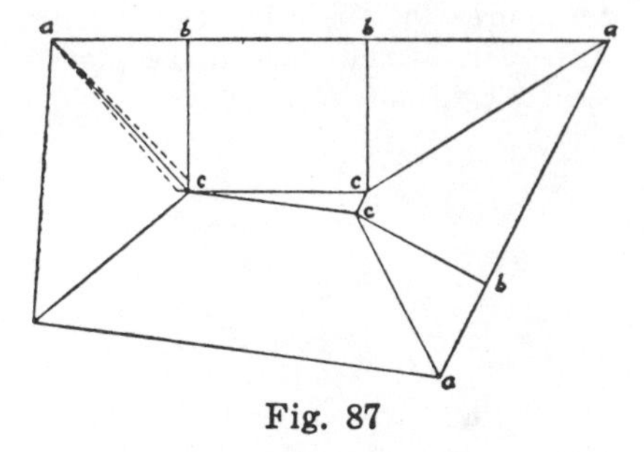

Fig. 87

the run and lying flat on one side. In other words, point *b* would have to be swung around until it would be directly above *c* in order to bring the rafter into its proper position. Only one hip is shown at the left, because the two left hips are the same. The shaded bevel at *a* gives the foot cut, while the one at *b* gives the plumb cut on each of the rafters shown.

Obtaining Figures for Cuts.—Fig. 85 is a diagram showing the four rafters just explained, the letters having the same significance. It will be noticed that the four rafters have the same rise, but each one has a different run. To the left is shown a square, in part, applied to the diagram in order to obtain the figures to be used when making the different cuts. The base figure, 12, is brought to the toes of the rafters and the point where the lines of the different rafters cross the right edge of the tongue gives, respectively, the different figures to be used on the tongue. The same diagram is shown in Fig. 86. Here the figure 8 is used on the tongue in each case, as indicated. The square in position *A*, shows 24 and 8 as the figures to be used to obtain the cuts. When the square is slipped back to position *B*, shown by dotted lines, 8 will again be the figure to be used on the tongue, but the figure to be used on the body of the square must be obtained on the diagram at the toes of the rafters. To obtain the figures for the other two rafters the square would have to be moved back to positions *C* and *D*.

Edge Bevel of Jacks. — **Fig. 87** shows another diagram of the plan we have been using. The dotted lines to the upper left are on a 45-degree angle with the respective sides. The plumb and foot cuts of the common and jack rafters are regular. The edge bevels of the jacks are different for each of the three hip rafters shown. To obtain the edge bevel for the jacks in this roof, take the distance *a-b* and the length of the common rafter per the distance *b-c*, the latter gives the cut. The diagram is based on a run of 12 inches, or one foot, which keeps all figures within the length of the body of the square, but in case of a plan where the distance *a-b* would be more than 24 inches, then divide both figures by two and use the quotients on the square to obtain the cuts.

LESSON 13

JACK RAFTER CUTS AND BACKING FOR HIPS

Cuts for Jack Rafters.—In the last lesson we explained how to obtain the edge bevel of jack rafters of an irregular-pitch roof. Let us review this, using Fig. 88 as a diagram. The reference letters, however, are not the same. We are showing three common rafters in triangular form lying on one side, in which *b-c* represents the run; *c-a*, the rise; and *a-b*, the rafter. Take the distance from *b* to the corner on one arm of the square, and *a-b* on the other, the latter gives the edge bevel for jack rafters in the corner under consideration.

Difference in Lengths of Irregular Jacks.—The next thing we want to know is the difference in the lengths of the jack rafters for each corner. On one side of each of the corners we have drawn in the jack rafters. Where these jack rafters intersect with the hip we show a dotted line running parallel with the plate to the common rafter, as from 1 to 1, 2 to 2,

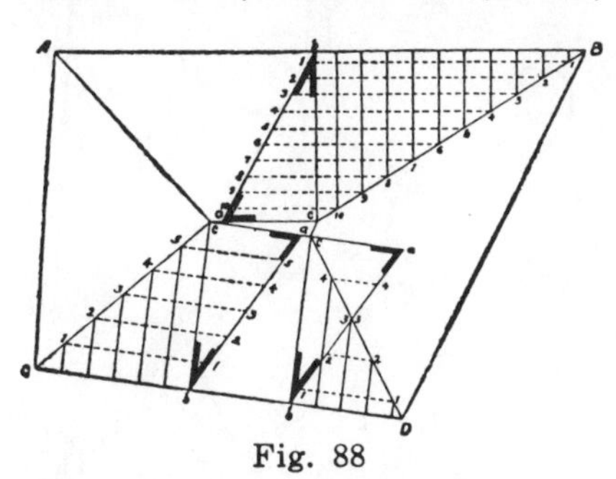

Fig. 88

3 to 3, and so on. The distance between any two points where these lines intersect the common rafter, gives the difference in the lengths of the jack rafters. In the corner marked *B*, it will be noticed, the first and the last spaces are shorter than the other spaces—the reason for this is that the spacing of the rafters does not work out so as to intersect with the corner of the deck or the corner of the building. The spacing of the other two rafters works out all right with the corners of the deck, but on

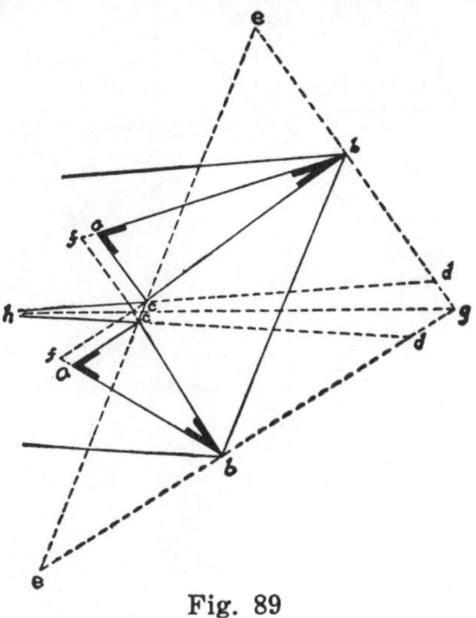

Fig. 89

one of them it does not work out with the corner of the building. The irregularity of the plan is responsible for this. The jacks for the hip marked *A*, are the same as the jacks for the hip marked *C*.

Edge Bevel for Hips. — Fig. 89 shows a part of the plan shown by Fig. 88, namely, the end to the right. This diagram shows the long and the short hips lying on one side—the rise, it will be noticed, is the same, but the run is different. What we want to know is how to get the edge bevel. To do this we proceed by drawing *e-e* parallel with *b-b* in line with the end of the deck, *c-c*. At a right angle to *b-c* of both the long hip and the short hip, draw *e-g*. Draw *h-g* and the two dotted lines from *c* to *d*, keeping them in line with the sides of the deck. Extend the line *b-a* to *f* on both rafters and draw the dotted lines parallel with *c-a*. This completes the diagram. We are assuming that the hips are to join the corner of the deck, which will require two edge bevels in order to saddle over the corner. Speaking of both rafters, the distance *e-b* taken on one arm of the square, and the distance *a-b* on the other will give the edge bevel that will fit the end of the deck, the distance *a-b* giving the cut. To get the bevel that will fit the side of the deck, take *b-d* on one arm of the square, and *a-b* on the other; the latter gives the cut. If the hip rafters are extended to *f*, and the deck is omitted, then the edge bevels for the two hips where they join each other are obtained by taking *b-g* on one arm of the square and *b-f* on the other; the latter gives the cut. The other edge bevels, in case the hips must fit into corners, are obtained just as in the other case, taking distance *e-b* and *a-b;* the latter gives the cut.

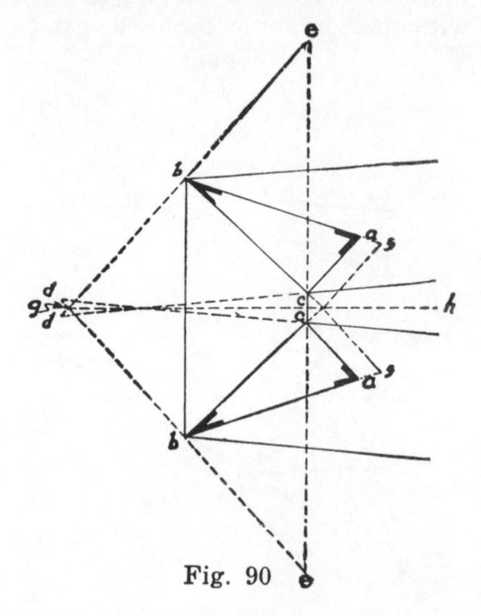

Fig. 90

Another Diagram Giving Cuts.—Fig. 90 is a diagram showing the narrow end of the deck, otherwise it is the same as Fig. 89. The reference letters are the same and the explanations of Fig. 89 will apply here. It should be noted, though, that the extended lines from the sides of the deck cross each other, whereas in the other diagram they flared out.

How to Get Bevel for Backing.—Sometimes it becomes necessary to back rafters. This is especially true when a blind valley rafter is used; that is, the valley rafter is extended beyond the valley in order to gain support. The part beyond the valley must be backed so as to keep the edge in alignment with the common rafters.

A method of obtaining the bevel for backing hips and valleys is shown in Fig. 91. The diagram shows how to obtain the bevel for backing hips.

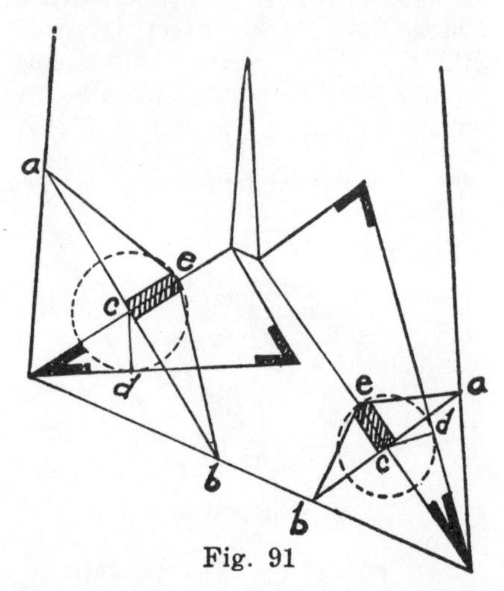

Fig. 91

The bevel for backing valleys is the same as for hips, but in reverse order. At a convenient point strike *a-b* at a right angle to the seat of the hip rafter. Where this line crosses the seat, strike *c-d* at a right angle to the line representing the rafter. Then strike the dotted circle and draw *a-e* and *e-b*, which give the bevel for the backing at *e*. The shaded part represents a section of the backed rafter.

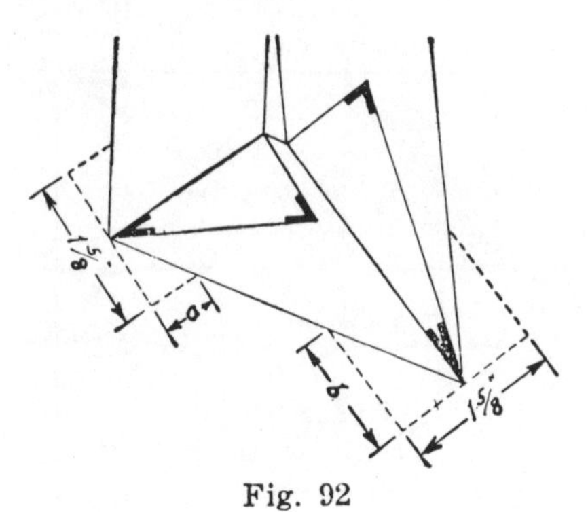

Fig. 92

Another Method for Backing.—Figs. 92 and 93 give another method of obtaining the bevel for backing rafters. Fig. 92 shows the plan we have been using, and the dotted lines represent the points of the unbacked hip rafters, which are made of 1⅝-inch material. Turning to Fig. 93 we show to the right both the short and the long hip rafters of Fig. 92. The distances *a* and *b* are transferred as shown and the depths of the backings are obtained as indicated by dotted lines. To the left we show sections of the rafter material, in part, backed for the two hips. For the valleys the application is in reverse order.

We want to emphasize here the importance of visualizing these problems into, as it were, realities; which is to say, that the student, by the aid of his imagination, should be able to see the solutions of these problems, as if he were solving them in

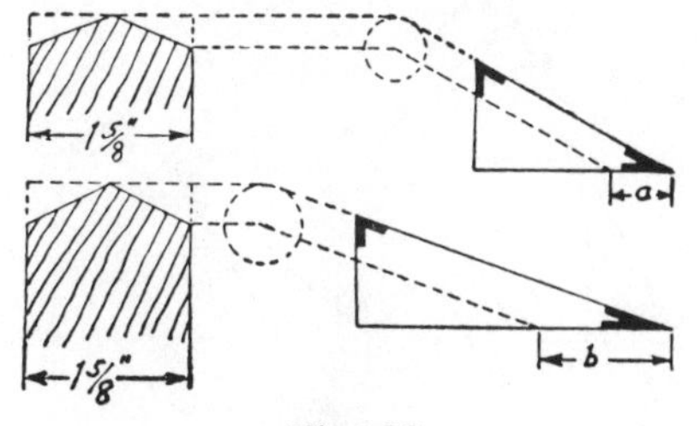

Fig. 93

actual practice. To cultivate the imagination to this degree, is an achievement that is more than an asset to the person who has accomplished it.

LESSON 14

IRREGULAR-PITCH ROOFS

Irregular Pitch.—In the irregular-pitch roof, or unequal-pitch roof, as it is sometimes called, the rise of all of the rafters is the same, but the rise per foot run is different. For example, if one is speaking from the standpoint of the rise per foot run, the difference in an unequal-pitch roof is always found in the rise; but, on the other hand, if one is speaking

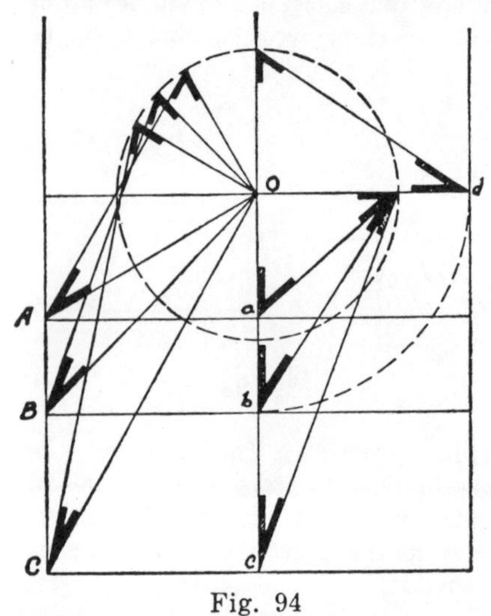

Fig. 94

from the standpoint of the full rise of the full run, the difference is always found in the run. These differences make it necessary for the roof framer to be constantly on guard, lest he apply the principle of one rafter to an entirely different rafter. When this happens undetected by the roof framer (who later finds it out) it can easily throw his mental conception of irregular-roof framing into confusion. But if one has a diagram before him, and understands it, it is possible to frame an unequal-pitch roof without confusion or other difficulties.

Rise and Run Compared.—**Fig. 94** is a diagram showing three hip rafters and four common rafters in tri-

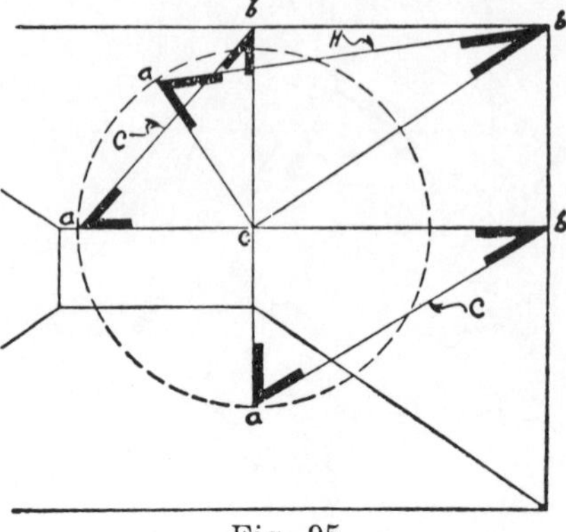

Fig. 95

angular form, lying on one side. This should be remembered in order to understand the drawing. Three different hip roofs are represented by this illustration—two are irregular and one is regular. The dotted cir-

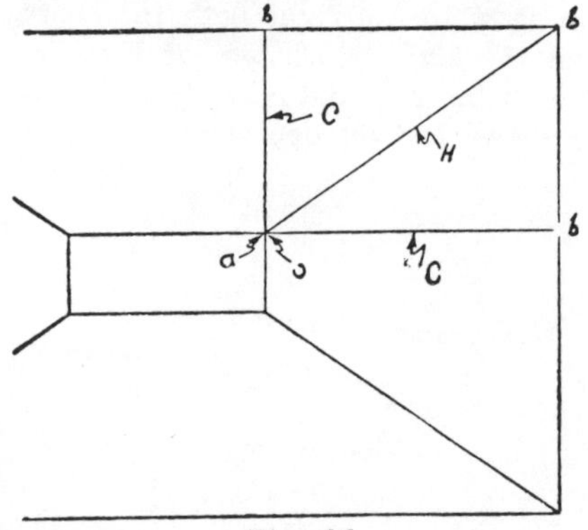

Fig. 96

cle, whose center is at *O*, shows that the rise of these rafters is equal, but it will be noticed that only two rafters have the same runs—they are shown from *d* to *O* and from *b* to *O*. The common rafters with the runs *c-O* and *a-O* are both for irregular roofs. The Hip rafter with the run, *C-O,* sets on a 60-degree angle; the one with the run, *B-O,* is on a 45-degree angle, and the one running from *A* to *O* is on a 30-degree angle. This diagram is given so that the student

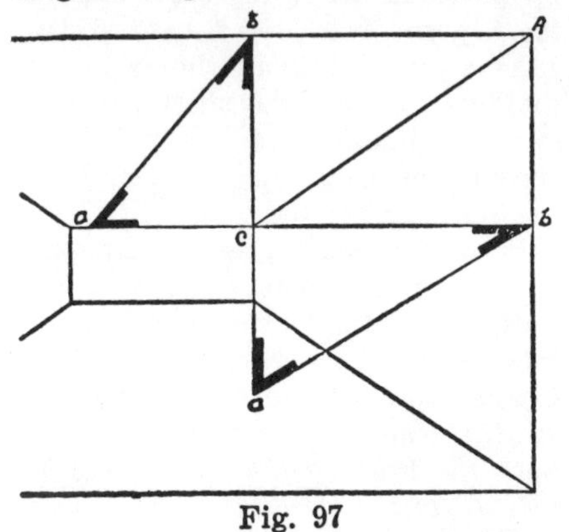

Fig. 97

can compare the different roofs represented and observe the rises and the runs.

Another Comparison.—Fig. 95 is a diagram showing two common rafters, marked *C,* and one hip rafter,

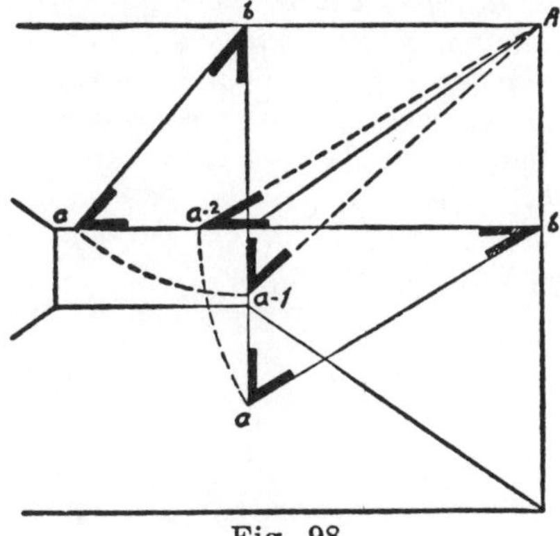

Fig. 98

marked *H,* lying on the side. In all of these rafters *b-c* represents the run, *a-b* the rafter, and *c-a* the rise. Now, if the rafters were raised into position, bringing point *a* directly above point *c,* the drawing in plan would appear like what we are showing by Fig. 96. Compare the two diagrams.

In the next three diagrams we take up obtaining the edge bevel of jack rafters for the irregular-pitch roof. We treated this under the irregular plan, but are bringing it in here again to show that the principles are the same in both cases, but the conditions are different.

How to Obtain Edge Bevel.—If the student will observe Fig. 96 and imagine laying the two common rafters back on the side, he will have a good idea of Fig. 97. To obtain the edge bevels of the jack rafters that join the hip, shown in position from *A* to *c,* take the distance from *A* to *b* on

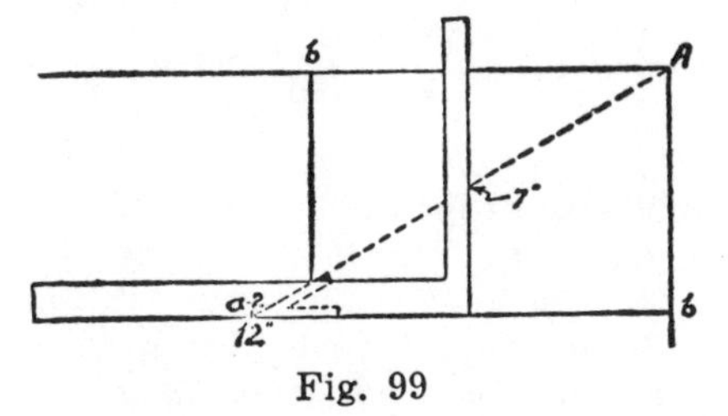

Fig. 99

one arm of the square, and the distance from *b* to *a* on the other arm; the latter gives the cut. Here it will be noticed that there is a difference in the edge bevels of the jacks that join the hip on one side and the jacks that join it on the other side. In the irregular-plan roof the edge bevels for jack rafters were the same on both sides of the hip.

Advanced One Step.—Fig. 98 shows the same drawing but carried one step farther. Here the lengths of the rafters, *a-b,* have been projected, as shown by the part-circles, until they are in the positions *b* to *a-1* and *b* to *a-2.* The shaded bevels shown at *a-1* and *a-2* are the respective edge bevels of the jack rafters. How to apply the square to the diagram in order to obtain the figures to use on the square is shown by Figs. 99 and 100, which show the two advanced steps of the diagram.

Fig. 99 shows the square applied to the diagram to obtain the figure to be used on the tongue with the base figure, 12, on the body. These figures, 12 and 7, will give the edge bevel

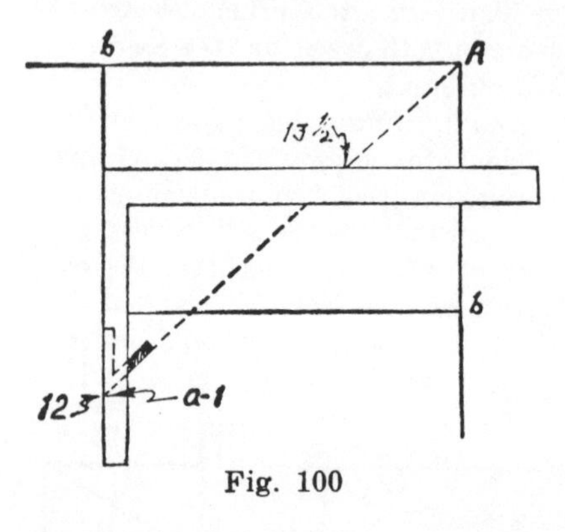

Fig. 100

of the jacks for the long run, 12 giving the cut. Fig. 100 shows the figures to be used on the square, 12 and 13½, to obtain the edge bevel of the jacks for the short run, 12 giving the cut.

We are recommending the use of diagrams, drawn to a convenient scale, because irregular-pitch roofs are not used enough for the workman to remember all of the rules, but if any one wants to frame an irregular-pitch hip roof he can obtain the edge bevels of the jack rafters by the following rules:

The length of the common rafter of the short run and the long run, taken on the square, will give the edge bevel of the jack rafters for the short run; the length of the rafter gives the cut. For the other side of the hip the rule would be just the reverse—the length of the rafter of the long run and the short run will give the edge bevel of the jack rafters for the long run; the length of the rafter gives the cut.

Here is a short rule for framing irregular-pitch roofs: *The run for valleys or hips is the diagonal distance between the long and short runs, which is used with the full rise in stepping off the length of hips and valleys. Both the run and the rise should be reduced to a scale of one inch equals one foot, and then take twelve steps. . . . The edge bevel for hips and valleys is obtained by taking the length of the hip or valley and the tangent on the square, the length of the rafter giving the bevel. (The tangent is the right angle distance from the toe of the hip or valley to the center of the building, or its equivalent.) For the edge bevel of jacks, take the length of the common rafter of one run and the other run on the square, the length of the rafter giving the bevel. The plumb and level cuts are the same as for the common rafters.*

LESSON 15

FRAMING IRREGULAR-PITCH ROOFS

Plan of Irregular-Pitch Roof.—In the last lesson we dealt with the edge bevels for jack rafters of irregular-pitch roofs, but in practice, before the edge bevel for the jack rafters is needed the framing of the hip rafters must be done and the lengths of the common rafters must be known. These things we will cover in this lesson. The plan shown by Fig. 101 represents an irregular-pitch roof with a long run of 18 feet and a short run of 12 feet. The deck, as the figures show, is 2 feet by 4 feet. In the

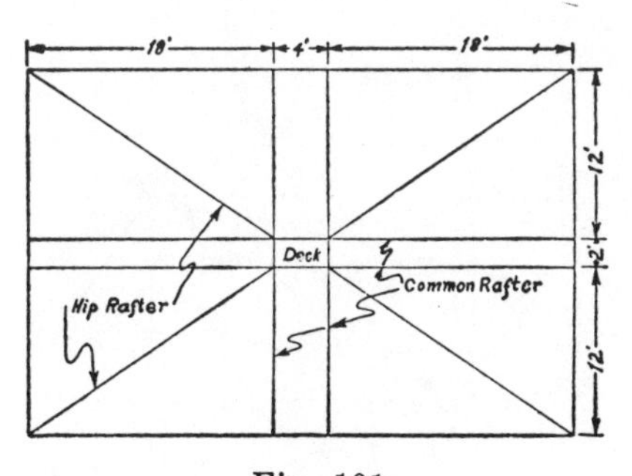

Fig. 101

illustrations that follow, the inches on the square will represent feet.

Run of Hip and Length of Hip Rafter.—Fig. 102 shows how to obtain the run of the hips for the roof shown by Fig. 101, which would be the diagonal distance in inches between 12, the short run, and 18, the long run, or $21\frac{7\frac{1}{2}}{12}$ inches, and would in reality mean 21 feet, 7½ inches. Having the run of the hip rafters we would obtain the length of the hip rafters in the manner shown by Fig. 103, where we take the diagonal distance between 8, the rise, and $21\frac{7\frac{1}{2}}{12}$, the run, which would give us $23\frac{1}{16}$ inches, or 23 feet, ¾ inch as the length

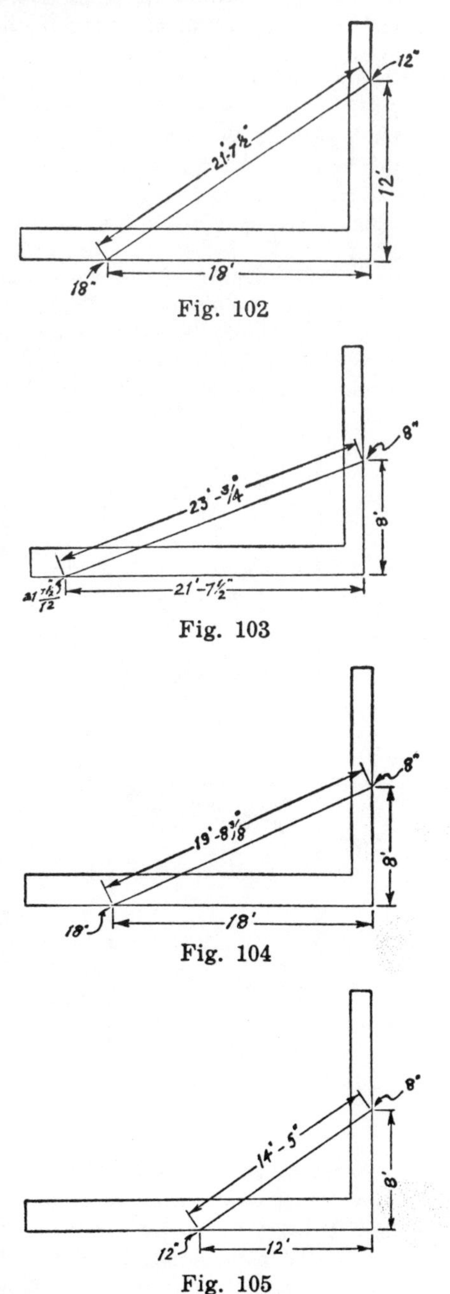

Fig. 102

Fig. 103

Fig. 104

Fig. 105

of the rafter. Now, $23\frac{1}{16}$ and 8, taken on the square, will give, respectively, the foot and the plumb cuts of the hip rafters.

Common Rafters of Long and Short Runs.—Fig. 104 shows how to obtain the length of the common rafter for the long run by taking the diagonal distance of 18 and 8 on the square, which gives us $19\frac{3}{4}$ inches, scant, or 19 feet $8\frac{3}{8}$ inches as the rafter length. The length of the short common rafter is obtained by taking the diagonal distance between 12 and 8, giving us $14\frac{5}{12}$ inches, or 14 feet 5 inches, as we are showing by Fig. 105.

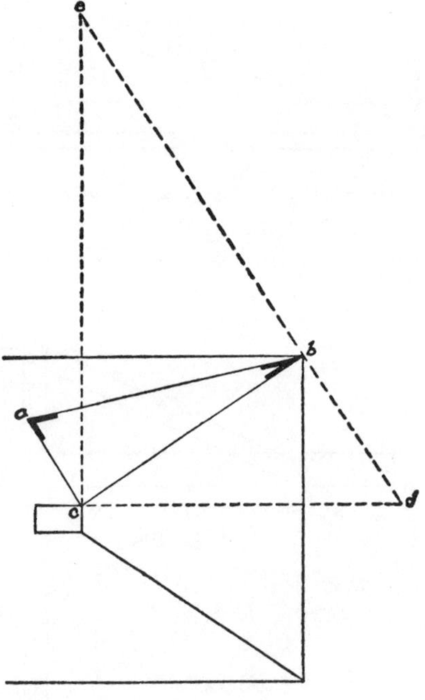

Fig. 106

Edge Bevel for Hips.—The next three illustrations show how to obtain the edge bevels for the hip rafters of the plan shown by Fig. 101. We will proceed by drawing the plan, in part, as shown, Fig. 106, and then lay off the triangle *a-b-c*, which represents the hip rafter. At a right angle to *b-c*, draw the dotted line *d-e*. Draw *e-c* parallel with the short run, and *d-c* parallel with the long run. The distance *a-b* and the distance *b-d*, taken on the square, will give the edge bevel that will fit the side of the deck; *a-b* gives the cut. The edge bevel that will fit the end

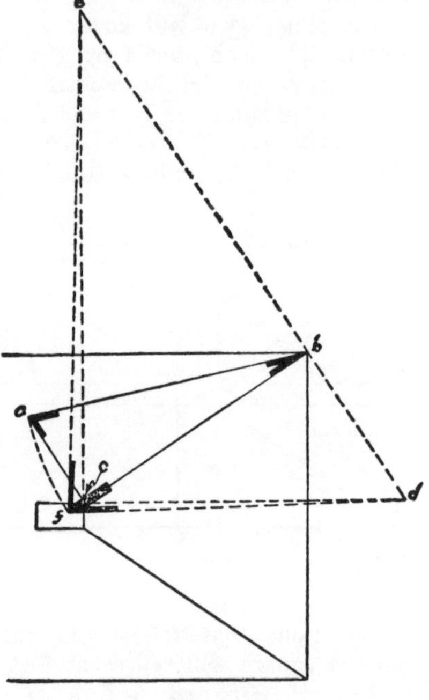

Fig. 107

of the deck is obtained by taking *a-b* and *e-b* on the square; *a-b* gives the cut.

Application of Square.—Fig. 107 shows the same diagram carried one step farther. Here the length of the rafter has been projected by dotted line from *a* to *f*, and this point has been connected with *e* and *d* by dotted lines. The bevel produced by the triangle *b-f-d*, gives the cut that will fit the side of the deck. The bevel produced by *b-f-e*, gives the cut that will fit the end of the deck. These two triangles, stripped of all unnecessary lines, are shown by Fig. 108 with a square applied to each of them. The base figure, 12, in each case, taken

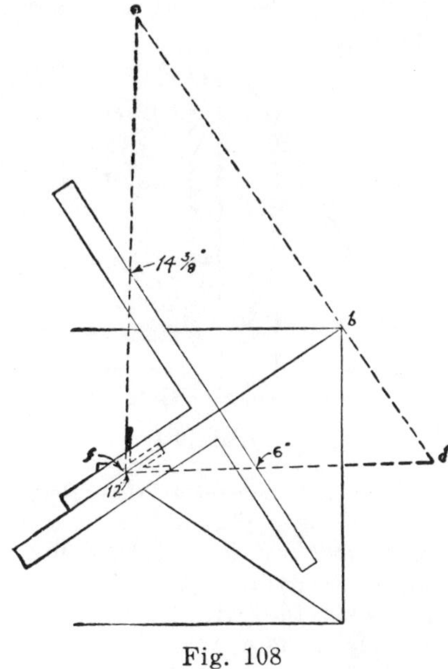

Fig. 108

on the square with the other figure, will give the cut. The cut fitting the side of the deck is obtained with 12 and 6, and the bevel that will fit the end of the deck is laid off by using 12 and 14⅜ on the square.

Stick to Steel Square.—I am aware that all of the cuts necessary to frame an irregular-pitch roof, or any other kind of roof, can be obtained by making use of higher mathematics. But for the average carpenter this is impractical, for it would mean, in most cases, a waste of time and energy. This being true, I conclude that the average carpenter should stick to the practical steel-square method of obtaining the cuts for any kind of roof framing.

While there might be available roof framing tables that will give the cuts and length of rafters for irregular-pitch and irregular-plan roofs, I have never seen such a work. If there is one on the market I am ready to add a copy of it to my library.

I have said nothing about the edge bevel for valley rafters, because it is commonly understood that the cuts for valleys, so far as the bevel is concerned, are the same as for hips.

LESSON 16

SHEETING CUTS FOR IRREGULAR-PITCH ROOFS

Cuts for Sheeting on Hip Roofs.—Before we take up the illustrations of this lesson, which deal with the cuts for roof sheeting on irregular hip roofs, we want to explain how to obtain the cuts of sheeting for regular hip roofs.

To obtain the face cut of sheeting for regular hip roofs, take the length of the common rafter on one arm of the square, and the run on the other, the run giving the cut. This can also

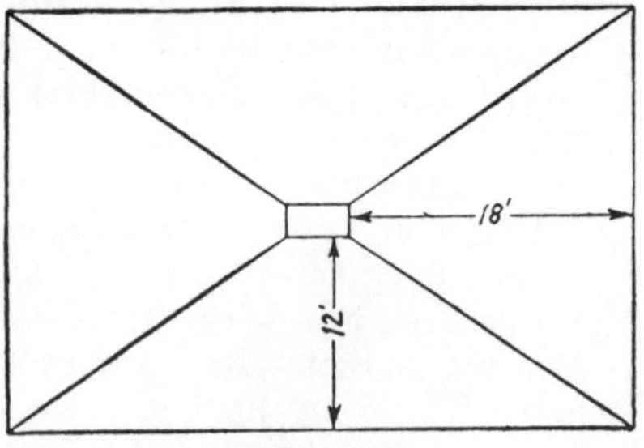

Fig. 109

be stated in this way: The length of the common rafter per foot run and the base figure, 12, taken on the square will give the face cut for sheeting, the base figure giving the cut. In obtaining the face cut for the sheeting, the figures to be used are the same as those used in getting the edge bevel for the jack rafters, excepting that the application of the square is in reverse order; which is to say, that the arm that gives the cut for the jack rafter does not give the cut for the sheeting, but the other arm does.

Edge Bevel for Hip Roof Sheeting.—To obtain the edge bevel of sheeting for regular hip roofs take the length of the common rafter and the rise of the roof on the square; the rise, in this case, will give the bevel. The same results can be obtained by taking the length of the common rafter per foot run and the rise per foot run on the square; the rise gives the bevel.

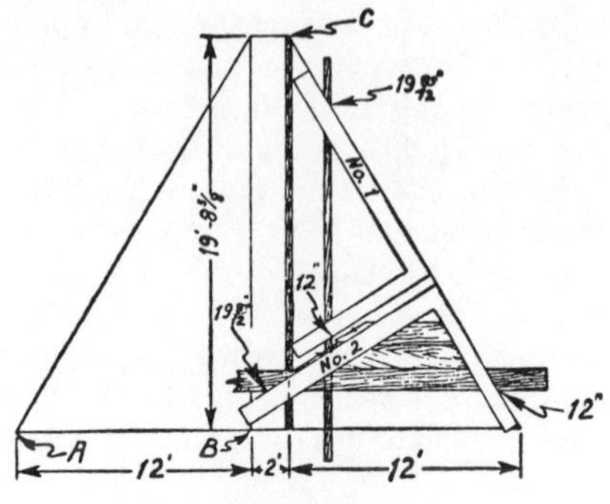

Fig. 110

Plan for Irregular-Pitch Roof.—Fig. 109 is a plan of the irregular-pitch roof we have been using in these lessons, and for which we are to obtain the cuts of the sheeting. The short run of this roof is 12 feet, while the long run is 18 feet.

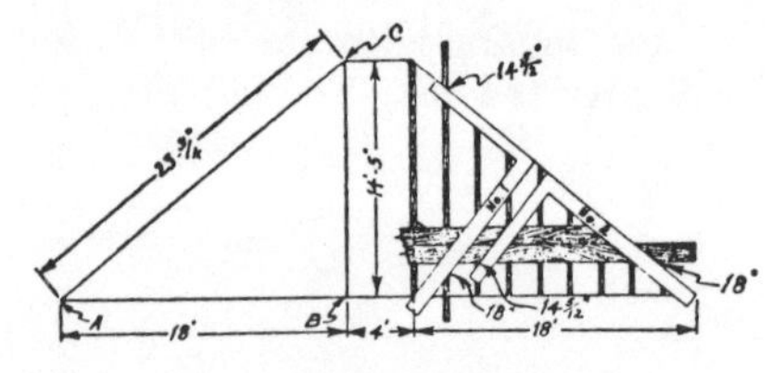

Fig. 111

Sheeting Cuts for End of Roof.—Fig. 110 shows one end section of this roof, looking straight at the surface. From *A* to *B* we have the distance of the short run, or 12 feet, which is represented as inches on the squares shown applied to the timbers. The deck is shown as being 2 feet wide, which has nothing to do with framing the roof, excepting that the rafters join it. The length of the common rafter for the long run is 19 feet 8⅜ inches. Now to obtain the edge bevel for the jack rafter we would take the short run, 12, and the length of the common rafter of the long run, $19\frac{8}{12}$, plus, on the square, the length of the rafter giving the cut. See the application of square No. 1. To mark the face cut of the sheeting, the same figures are used, but the square is applied in reverse order, as shown by square No. 2. If the student will keep in mind the fact that the edge bevel of the jack rafter and the face cut of the sheeting are obtained with the same figures, but applied in reverse order, he will not have any trouble in making these cuts.

Sheeting Cuts for Side of Roof.—Fig. 111 shows a section of the short run of the roof, looking straight at the surface. Here the long run is shown between *A* and *B*, which is 18 feet. The deck is shown as being 4 feet long. The common rafter for the short run is 14 feet 5 inches long. Now, to obtain the edge bevel of the jack rafter for the short run, take the length of the common rafter, or $14\frac{5}{12}$, on one arm

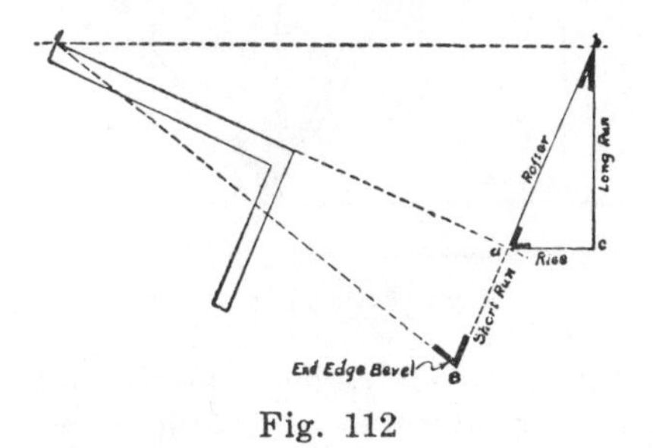

Fig. 112

of the square, and the long run on the other, the length of the rafter giving the cut, as shown by the application of square No. 1. To obtain the face cut for the sheeting, the square is applied in reverse order, as shown by square No. 2. A little study of Figs. 110 and 111 will show that there isn't anything complicated or difficult about getting these cuts, in fact, the principle is the same as in regular-pitch hip roofs.

Edge Bevel for Irregular-Pitch Sheeting.—In practice the edge bevel of sheeting is usually made without marking, but that the student might

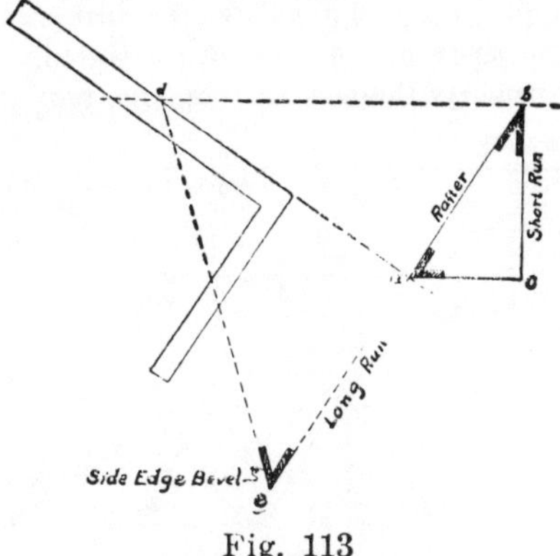

Fig. 113

know how to obtain it, we are showing two diagrams, one for the long run and another for the short run. Fig. 112 takes up the long run. The triangle *a-b-c* represents a common rafter for the long run, lying on one side. Proceed by drawing *b-d* at a right angle to *b-c*. Then draw *a-d* at a right angle to *a-b*. Extend *b-a* to *e*, making the extension equal to the short run, and draw *e-d*. The application of the square to the triangle *a-d-e* shows how to obtain the edge bevel of sheeting for the long run.

Fig. 113 shows how to obtain the edge bevel of sheeting for the short run. The triangle *a-b-c* represents a short-run rafter. Now draw *b-d* at a right angle to *b-c*, and *a-d* at a right angle to *a-b*. Extend *b-a* to *e*, making the extension equal to the long run. Draw *e-d*, which gives the edge bevel of the sheeting for the short run. The application of the square as shown on the diagram, shows how to get the figures to be used.

LESSON 17

HIPS, VALLEYS AND JACKS FOR IRREGULAR PITCH

Similarity in Framing Hips and Valleys.—In regular-pitch roofs valley and hip rafters are framed exactly the same, so far as the length and the various cuts are concerned; however, where hips and valleys are backed, which is seldom done, the backing for the valleys is just the reverse from the backing for the hips. In theory, though, rafters are framed

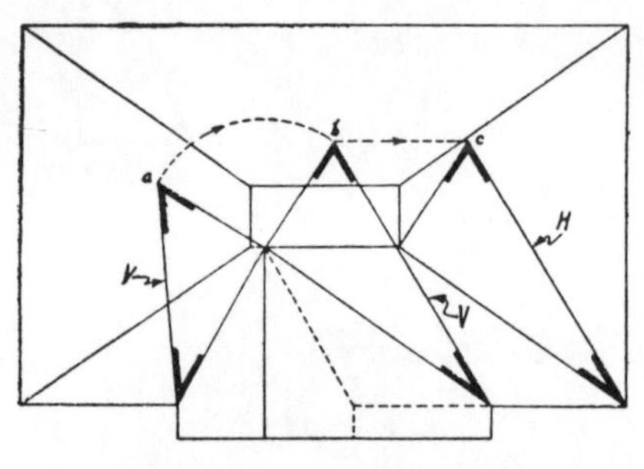

Fig. 114

with the center line as a base to work from; and so long as the roof framer is working to the center line in laying off hips and valleys, the framing

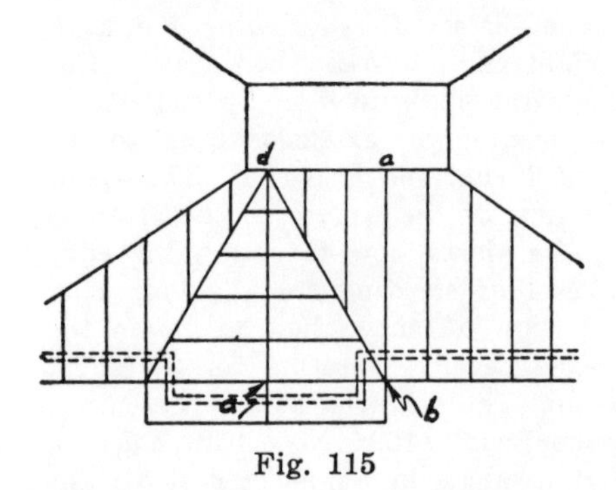

Fig. 115

of one is the same as the framing of the other. That is to say, the principle is the same; and in cases where everything else is equal, the rafters, after they are framed, will practically be the same.

Diagram of Valleys and Hips.—Fig. 114 is a diagram of an irregular-pitch roof, showing two valley rafters, marked *V*, and one hip rafter, marked *H*, in triangular form, lying on one side. At *a* is shown shaded, the top bevel for the short valley. The rise of this valley rafter is transferred by dotted line to the long valley, whose top bevel is shown at point *b*. The same rise is then carried to the hip rafter where the top bevel is shown at *c*. It will be noticed that the long valley and the hip rafters shown are exactly the same in run, rise, rafters and bevels. But if the roof were to be framed on the basis of the dotted lines shown to the bottom, then the two valley rafters would be the same in every re-

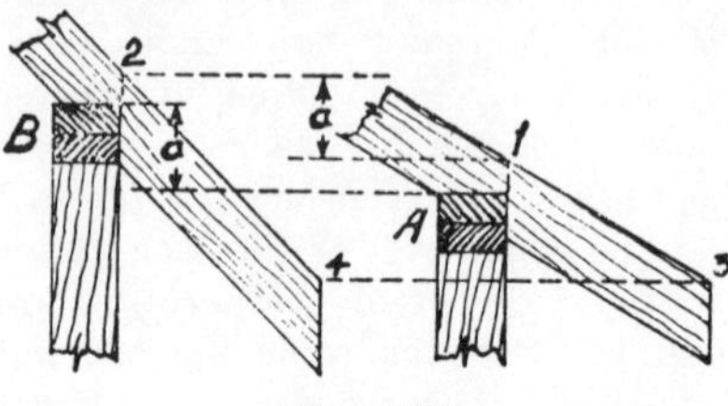

Fig. 116

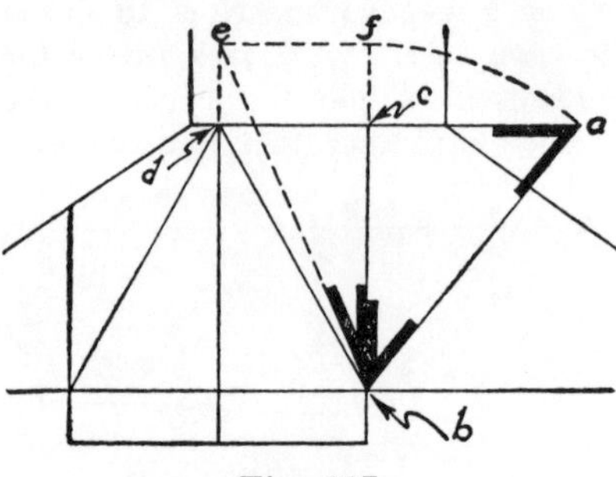

Fig. 117

spect, excepting that they would have to be framed in reverse order. This layout is given again by Fig. 115, where the rafters are shown by single lines.

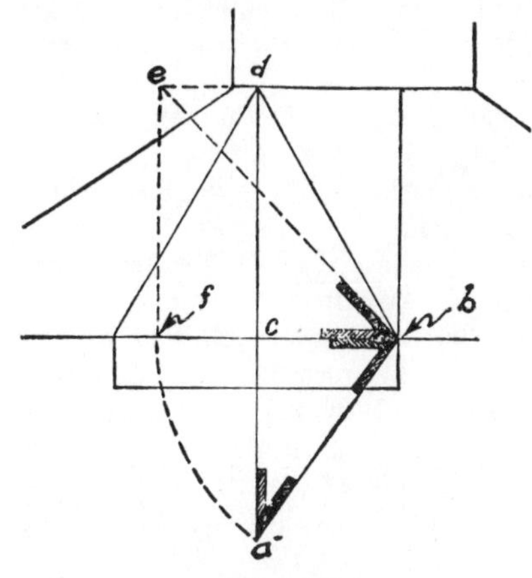

Fig. 118

Cuts for Jacks.—To obtain the edge bevel for the valley jacks, take the length of the rafter for the run *a-b*, Fig. 115, and the distance *a-d* on the square; the length of the rafter gives the bevel. We are using the same reference letters for both sides of the valley rafter, which makes the procedure the same for obtaining the bevels for the jacks of both sides of the valley.

Irregularity of Plates.—Up to this point we have been assuming that the rafters did not extend beyond the plate line. The dotted lines in Fig. 115 show what would happen if the rafters had to extend beyond the plate. The valley rafter would not rest on the plate directly over the angle, as in common roof framing, but would come to one side as shown. A similar situation would exist in cases of hip rafters, speaking of irregular-pitch roofs. But the matter isn't as simple as that, for the plate carrying the steeper of the two pitches would have to extend beyond the one for the low pitch. Fig. 116 illustrates this. Here *A* shows the low pitch and *B* the high pitch. The difference between point 1 and point 2, represented by the letter, *a*, gives the difference in the height of the two plates. Point 3 and point 4 must be on a level. This illustration should be studied on the basis of the dotted lines shown in Fig. 115.

Further Developed Diagram. — Figs. 117 and 118 show how to obtain the edge bevels of the valley jacks carried one step farther than what is shown in Fig. 115. Here the triangles *a-b-c* represent the common rafters lying on the side. With a compass set at *b*, transfer *a* to *f*, in both diagrams. Then extend *b-c* to *f* and strike *d-e*, making it parallel with *c-f*. To obtain the edge bevel, take *b-f* on one arm of the square, and *e-f* on the other, the former giving the bevel. The same explanation applies to the diagram shown by Fig. 118, where the reference letters are the

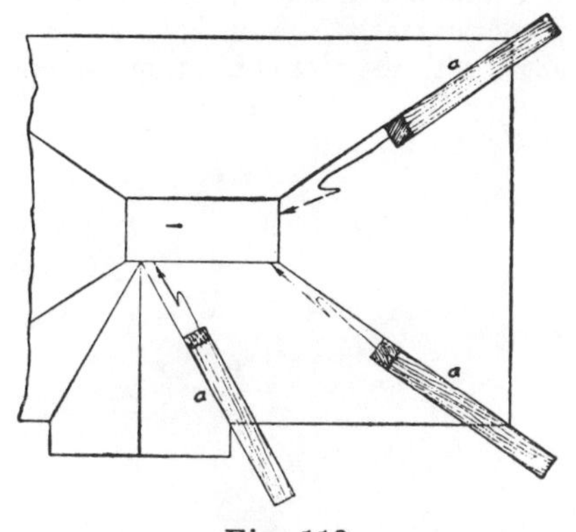

Fig. 119

same, but the bevel obtained is for the other side of the valley rafter. The plumb cut for valley jacks is the same as for the common rafters of the respective sides.

Practical Way to Obtain Bevels.—Fig. 119 illustrates a very practical way of obtaining the edge bevels for hip, valley and jack rafters for regular- and irregular-pitch roofs. Let us take up obtaining the edge bevel for hips and valleys first: Take a piece of rafter material and give it the foot cut on the order shown by Fig. 120. Then set the piece on the plate as shown in Fig. 119, *a*, *a*, *a* and

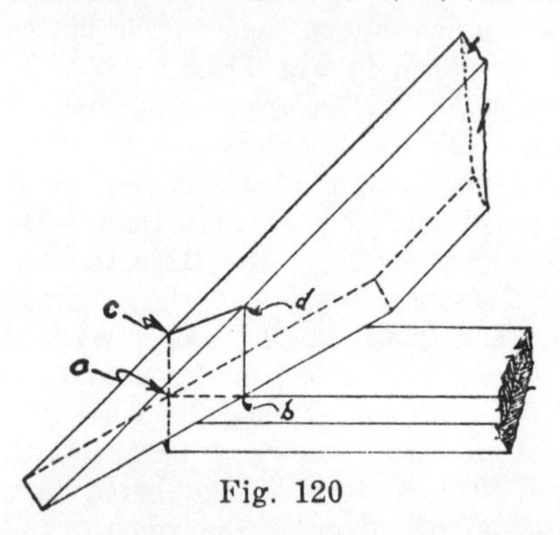

Fig. 120

mark the bottom with a pencil as shown by dotted lines. Now turn to Fig. 120, where we have a perspective view of the piece with the foot cut. The dotted line *a-b* shows where the first mark is made. At a right angle with the foot cut, strike *a-c* and *b-d* and connect *d-c*—*d-c* is the edge bevel of the hip or valley rafters, respectively. This cut will fit the deck as pointed out with arrows in Fig. 119. It also gives the edge bevel that will fit the plate on the side where the block is set for marking.

The procedure is the same for obtaining the edge bevels for hip or valley jacks, with two exceptions. First, you must use the foot cut of the common rafter, and mark it the same as explained for hips and valleys, but the bevel obtained on one side of a hip or valley line, as shown in Fig. 119, will give the bevel in reverse order for the jacks of the opposite side of the hip or valley, respectively. In other words, if you want the bevel for the jacks that will fit the hip or valley rafters on the side marked *a*, set the block where it is shown in Fig. 119; but if you want the bevels for the side where the blocks are shown, set the blocks on the side marked *a*, and mark them. These illustrations should be studied and tested in actual practice.

LESSON 18

TRIANGLE AND SQUARE

Roof Framing a Test for Leadership.—Perhaps there is nothing else coming under the head of carpentry, that tests men's ability for leadership, as roof framing does. A man may be the leader on a job, but if he

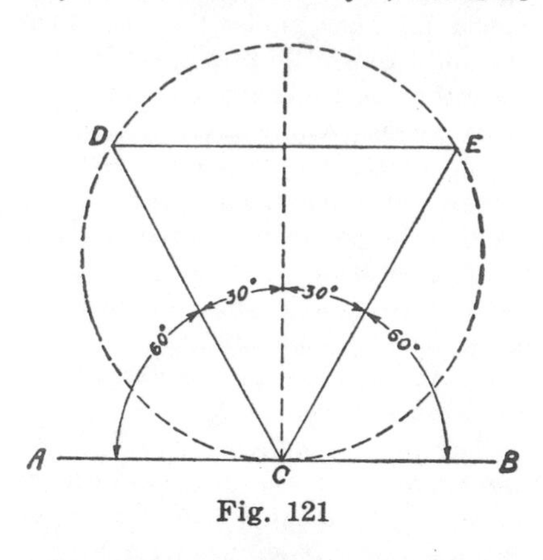

Fig. 121

fails when it comes to framing the roof, another man is called in to lead, because the man who can frame a roof, or any kind of roof, is better fitted to be the leader of a gang of

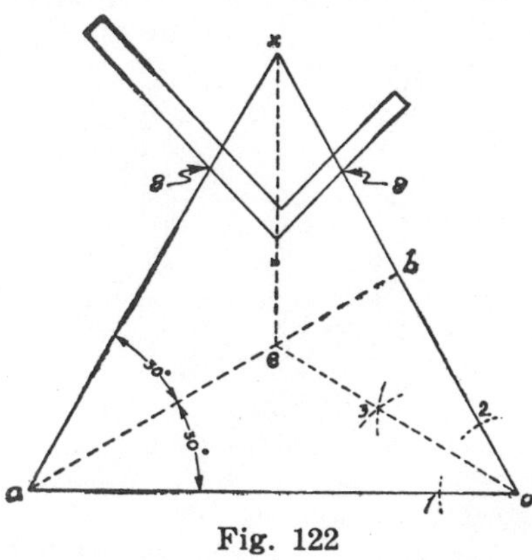

Fig. 122

builders than the man who can not master the simple principles of roof framing. I do not mean by this that if a foreman fails to be able to frame any kind of roof, that he will lose his job immediately, though that some-

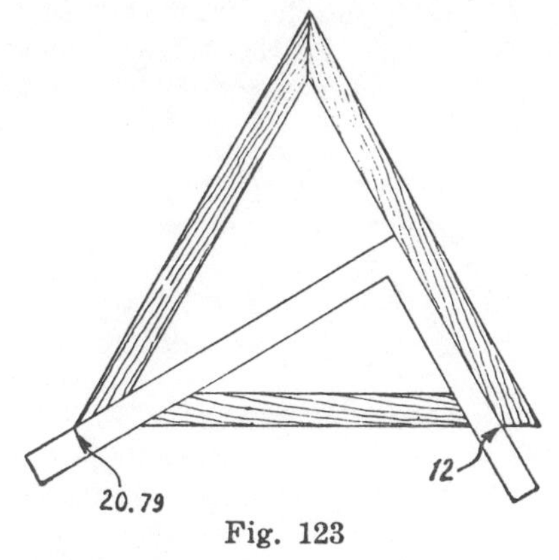

Fig. 123

times happens; but I do mean, that in due time that very thing will happen, and the man who is able to do

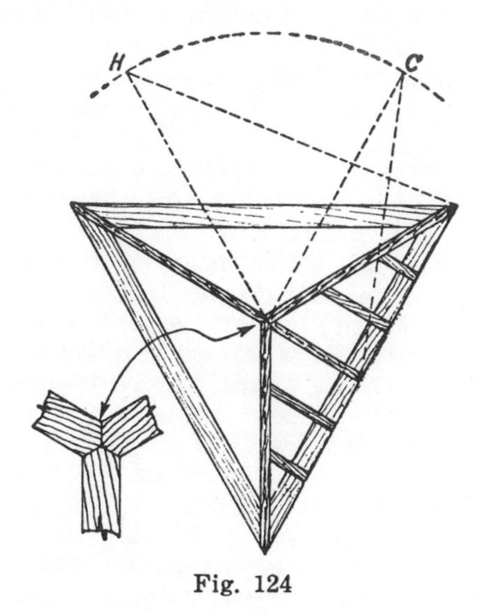

Fig. 124

any kind of roof framing will be the man who will take his place. This change might not come for weeks or months or even years after the failure, but eventually it must come, unless the man who made the failure, in the mean time qualifies himself to master roof framing.

The principles for framing a roof for a polygon are the same as those explained under irregular-plan hip roofs, excepting the quadrilateral, which would come under regular hip roof framing.

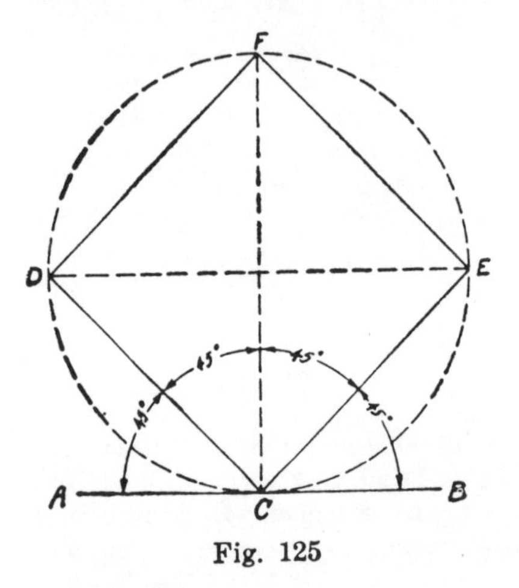

Fig. 125

Triangle.—Fig. 121 is a diagram showing how to lay off a triangle within a circle. If you have the length of one side of the triangle you want to describe, multiply it by .57735 and it will give you the radius of the circle to be used in laying off the triangle. After the circle has been struck, draw a line from *A* to *B*. From *C* draw *C-D* and *C-E* on a 60-degree angle. Then draw *D-E*, and your triangle is laid off.

Locating Center.—Fig. 122 shows four methods of locating the center of a triangle. First, strike a line 30 degrees from the sides of two of the angles and draw a line from each of the angles to a little past *c*, as shown from *a* to *c*. The crossing of the two lines at *c*, will establish the center of the triangle. Second, locate the center of two of the sides, as at *b*, and strike a line from each of these center points to the opposite angle; where the two lines cross is the center. Third, take two equal figures on the square and make them intersect the side lines of an angle an equal distance from the angle, as at 8 and 8. Then strike the line from the angle to a little past *c*—strike another line in the same way from another angle. Where these lines cross is the center point. Fourth, set the compass at *o* and strike 1 and 2. Then from these points strike the cross at 3. A line from *o*, crossing point 3, to a little past *c*, and another line from another angle, made in the same way, will locate the center where these lines cross.

Cuts for Triangle.—Fig. 123 shows what figures to use on a steel square in order to obtain both the miter cut and the butt-joint cut for a plate of a triangular building.

Rafters for Triangular Roof.—Fig. 124 shows a triangular plan of a roof with the hip rafters in place and also the jacks for one side. At *H* we are showing the top of the hip rafter in triangular form lying on one side, and at *C* we have the top of the common rafter. The dotted part-circle shows that the rise of the two rafters is the same. A detail of the top joint of the hip rafters is shown to the left, pointed out with indicators.

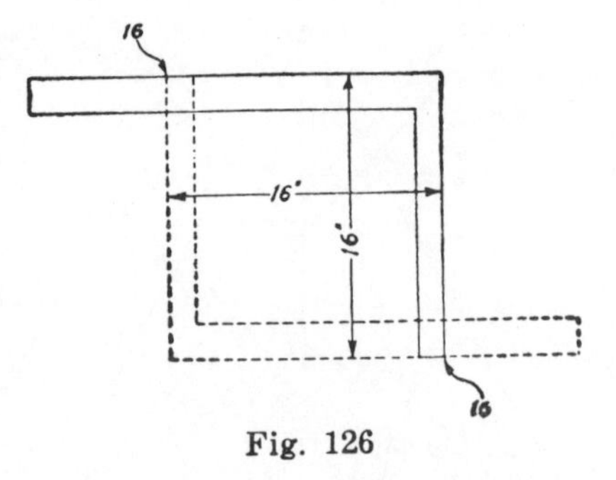

Fig. 126

Laying Off a Square.—Fig. 125 shows how to lay off a perfect square within a circle. If you know the length of one side of the square that you want to describe, multiply it by .70711 in order to obtain the radius

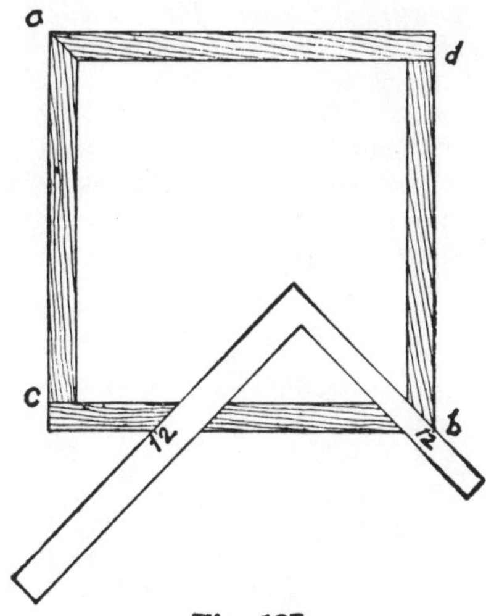

Fig. 127

for the circle. From line *A-B*, at *C*, strike *C-D* and *C-E* on a 45-degree angle, then strike *D-F* and *E-F*, also on a 45-degree angle, and you will have a perfect square. To find the center, strike *D-E* and *F-C*—where these lines cross is the center.

Steel Square Method.—Fig. 126 shows how to lay off a perfect square with the steel square by taking equal figures on the tongue and on the blade of the square; in this case, 16 and 16 are used, giving us a square 16″x16″.

Cuts for Joints.—Fig. 127 shows the square applied for obtaining the miter cut for the plate of a quadrilateral building. Here we are using base figures, 12 and 12, but any other equal figures will give the same results. The butt joint is made by cutting the timber square across.

Rafters for Roof.—A plan of a quadrilateral roof with the hips in place and also the rafters for one

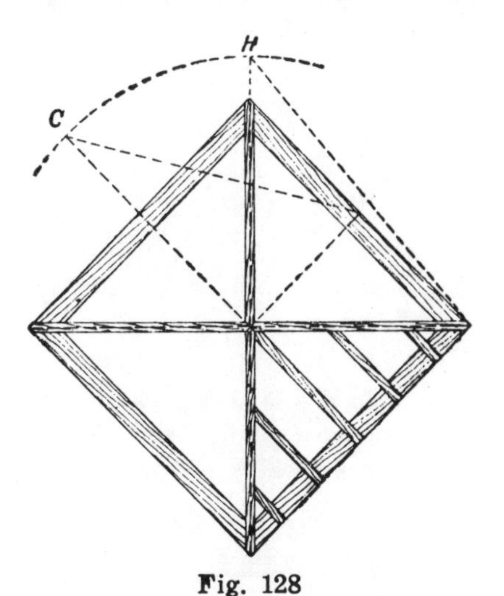

Fig. 128

side is shown by Fig. 128. The triangle at *C* represents the run, rise and the rafter of a common rafter lying on one side, while the triangle at *H* represents the hip rafter.

LESSON 19

MORE POLYGONS

Pentagon.—If you have the length of one side of a pentagon, multiply it by .85065 and it will give you the radius for a circle within which the pentagon you want can be described.

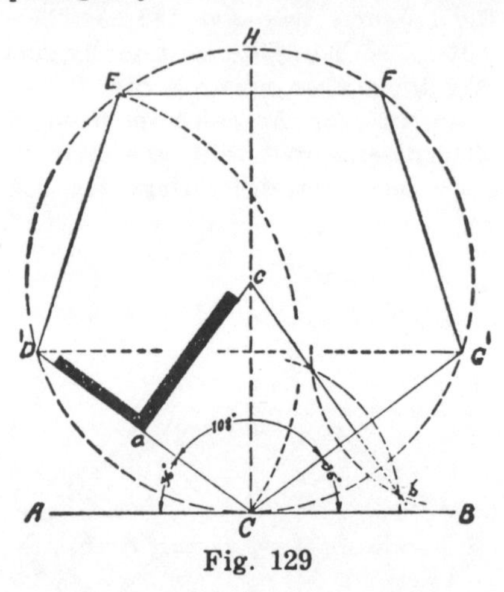

Fig. 129

How to do this is shown by Fig. 129. From line *A-B*, at *C*, strike *C-D* and *C-G* on a 36-degree angle. Then set

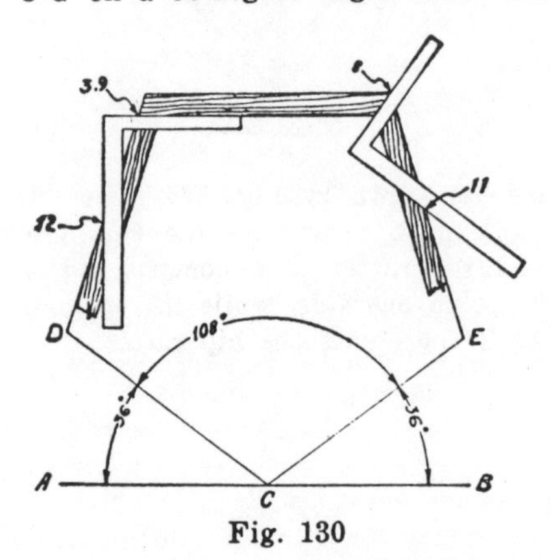

Fig. 130

your compass at *D* and strike the part-circle from *C* to *E*. Strike *E-F* parallel with *A-B*. Now strike *E-D* and *F-G*, and you have a perfect pentagon. Point *F* can also be established by setting the compass at *G* and striking a part-circle, just as we did in locating point *E*.

Locating Center.—To locate the center of a pentagon (the principle is the same in all polygons), strike *C-H* at a right angle with *A-B* or *E-F*. Then locate the center of one of the sides, say, *C-D*, which would be at *a*, and set the square at this point in the position shown and extend the

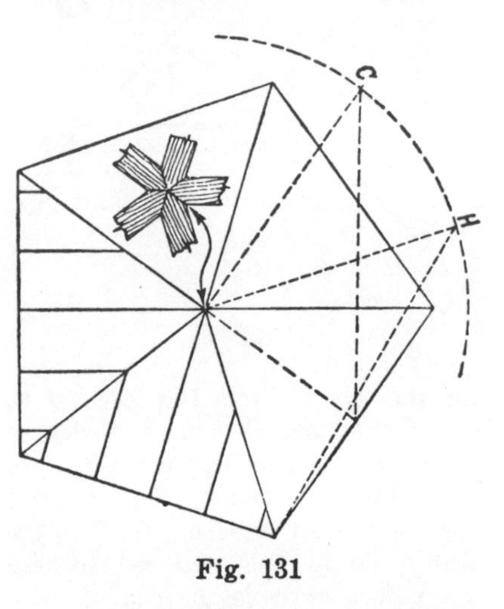

Fig. 131

line until it intersects line *C-H*, or point *c*, which gives you the center.

Another method is shown to the right. Set the compass at *C* and also at *G* and strike the two part-circles as shown, and draw the line from *b* to *c*, crossing the points where the part-circles cross. Where this line intersects with line *C-H* is the center of the pentagon.

Miter and Butt Joints.—Fig. 130 shows a plan of a pentagon showing what figures to use for obtaining the miter cut and the cut for a butt joint in framing the plates.

Rafters.—Fig. 131 is a one-line drawing of a plan for a pentagon roof. Here we show the rafters in place on two sides—we also show the run, rise and rafter in triangular form, lying on one side, of a common

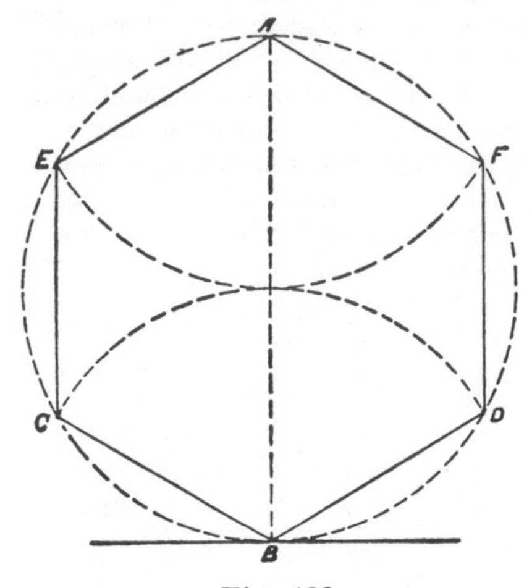

Fig. 132

rafter and of a hip rafter. Inset and pointed out with indicators, we show the top joint of the hip rafters in plan.

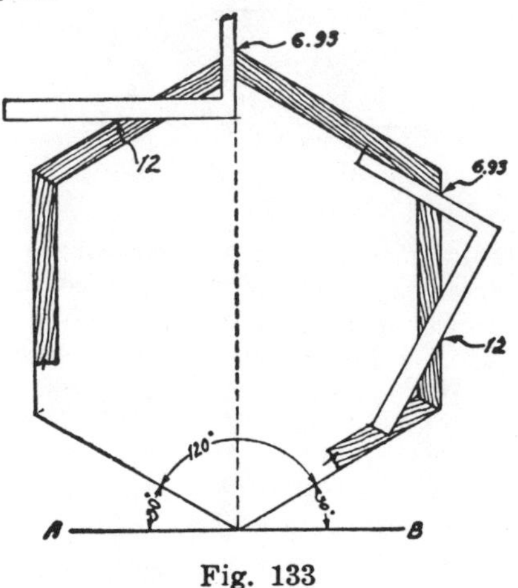

Fig. 133

Hexagon.—If you have the length of one side of a hexagon, and multiply it by 1, it will give you the length of the radius of a circle within which that particular hexagon can be described. In other words, if you have the length of one side of a hexagon you have the length of the radius of a circle within which that hexagon can be described.

Describing Hexagon. — Fig. 132 shows, perhaps, the simplest method of describing a hexagon: Strike the circle, and without changing the compass, strike the two part-circles from point *A* and point *B*, locating points *E*, *F*, *C* and *D*. Then draw in the sides as shown and you have a true hexagon.

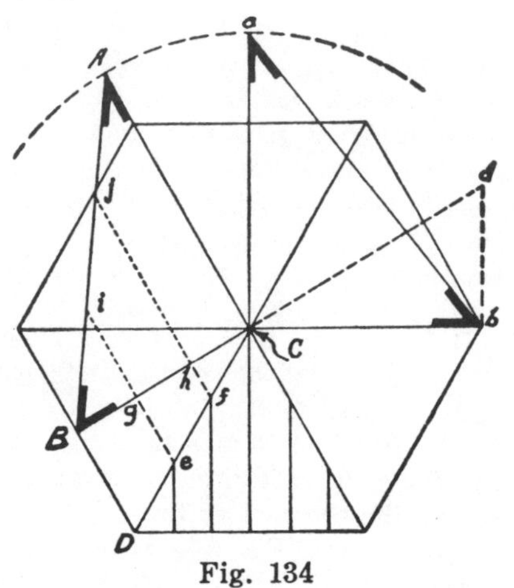

Fig. 134

Cuts for Hexagon.—Fig. 133 shows a plan of a hexagon with some of the plates in place. The squares show the figures to be used for obtaining both the miter cuts and the cuts for the butt joints. The degrees of the angles are also given, so that the student may describe this polygon within a circle in the same manner as explained for the previous polygons.

Rafters for Hexagon Roof.—A one-line drawing of a plan for a hexagon roof is given by Fig. 134. Here we show a common rafter, *A-B-C*, and a hip rafter, *a-b-C*, lying on one side.

To get the edge bevel for the hip rafter, take the length of the hip rafter, *a-b*, and the distance, *b-d*, on the square; the length of the rafter gives the cut. To obtain the edge bevel for the common rafter to fit into the angle at the top, take the length of the rafter, *A-B*, and *B-D*

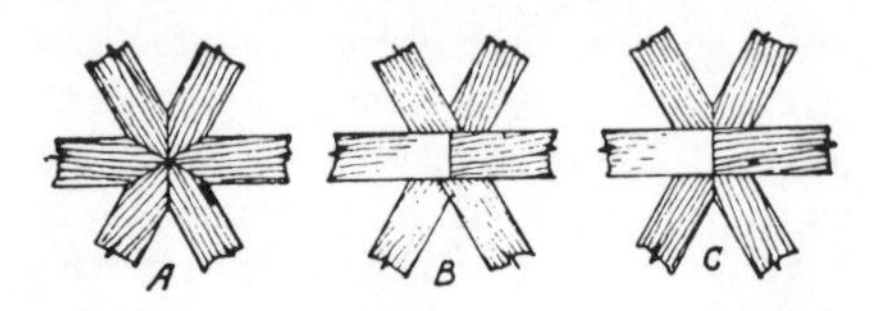

Fig. 135

on the square; the former gives the cut. This also gives the edge bevel for the jack rafters. In fact, the common rafter in polygon roof framing is a sort of jack rafter.

Difference in Lengths of Jacks.—How to get the difference in the length of the jack rafters is shown by the dotted lines: At a right angle with *B-C*, strike *f-h-j* and *e-g-i*. Now, *B-i* is the length of the shortest jack rafter, *B-j* is the length of the second jack rafter and *B-A* is the length of the full jack rafter, or what we are calling the common rafter.

Top Joints.—Fig. 135 shows three different top joints for the hip rafters of a hexagon roof. The one shown at *A* is probably the best of the three shown, while those shown at *B* and *C* might have some advantages; however, we believe that they are harder to frame. The last two joints can be modified so as to be suitable for other polygons.

Describing Heptagon and Cuts for Joints.—To obtain the radius of a circle within which a given heptagon can be described, multiply the length of one side of the heptagon by 1.15238 and the result will be the radius. After the circle is struck, set the compass to the length of one side of the heptagon and step off the seven sides on the circle. This done, draw in the lines from point to point until the heptagon is completed.

Fig. 136 shows a heptagon within a circle, giving the degrees and minutes of the angles. In describing a heptagon within a circle when the length of the sides is not known, proceed just the same as explained in previous polygons, excepting some

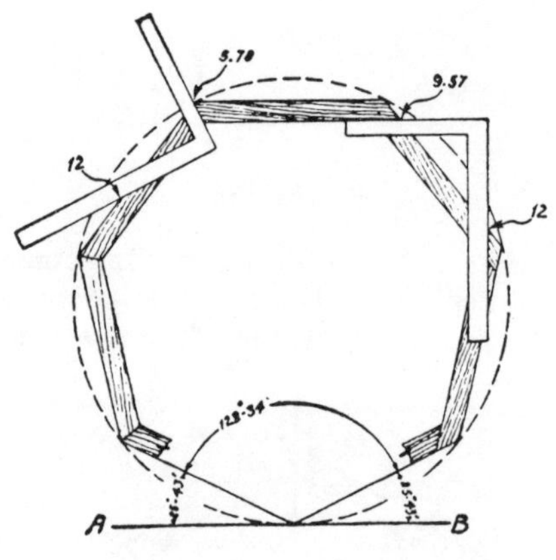

Fig. 136

adjustments that we expect the student to make. Two squares are shown applied, giving the figures to be used for making the miter joints of the plates and also the figures for butt joints.

LESSON 20

THE OCTAGON AND BACKING RAFTERS

The Octagon.—The octagon is probably the most used polygon of them all, especially in the line of carpentry. It is easy to describe, and lends itself to many situations where other polygons would hardly be practical.

If the length of one side of an

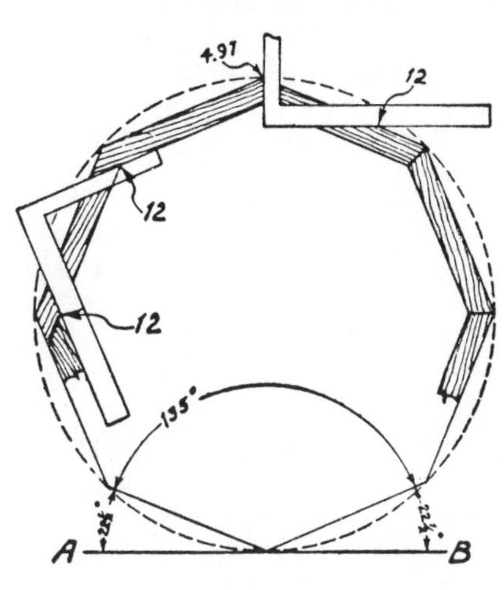

Fig. 137

octagon is known, multiply it by 1.30656 and you will have the length of the radius that will make the circle within which that octagon can be described.

Circle Method and Cuts.—Fig. 137 shows an octagon plan, giving the degrees of the angles so the student can lay off the octagon by the circle method if he cares to do so. We are also showing some of the plates in place with two squares applied giving the figures to be used for making the miter cut and the cut for butt joints.

Simple Method.—Fig. 138 gives a very simple method of describing an octagon within a circle. After striking the circle, strike *a-e* at a right angle with *A-B*, crossing the center of the circle. Then strike *g-c* parallel with *A-B*, also crossing the center of

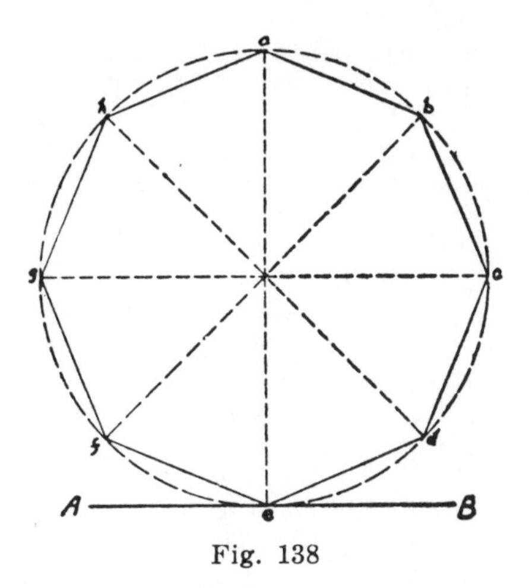

Fig. 138

the circle. This done, strike *f-b* and *h-d* on a 45-degree angle with *g-c* and crossing the center of the circle. Now

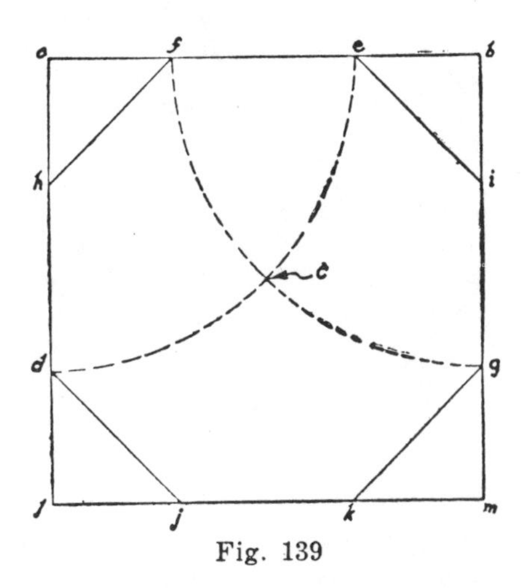

Fig. 139

draw in the sides of the octagon, as shown by the drawing, and your octagon is completed.

Describing Octagon within a Square.—Fig. 139 shows how to lay off an octagon by using the quadrilateral to start with. From points *a* and *b* strike the two part-circles in such a manner that they will cross the center, marked *c*, locating points *f*, *g*, *d* and *e*. On a 45-degree angle, draw *f-h*, *g-k*, *d-j* and *e-i*, and you have a perfect octagon. Sometimes two more part-circles are struck from points *l* and *m* in order to locate points *h*, *i*, *j* and *k*.

Backing for Hips.—Fig. 140 is a diagram of an octagon plan showing how to obtain the backing for the hip rafters, in case they have to be backed. The triangle *A-B-C* shows a hip rafter lying on one side. Draw *a-b* at a right angle with *A-B* and *b-c* at a right angle with *C-B*. Strike *d-e* parallel with *A-B*, and with the compass set from *b* to *c*, strike the half-circle from *d* to *e*. Now draw lines *a-d* and *a-e*, which give the backing bevels of the hip rafters at point *a*.

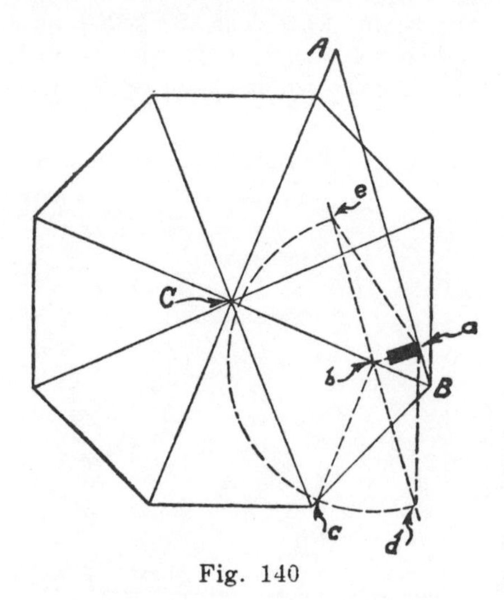

Fig. 140

Cuts for Sheeting.—Fig. 141 is a diagram of a plan, in part, of an octagon roof, bottom, and an elevation of the same roof above. This diagram shows how to get the bevels for the sheeting cuts. For the edge bevel, strike *b-c* at a right angle with *b-j*. Then make *a-b* equal to *d-e* and draw *a-c*. The bevel at *a* gives the edge cut of the sheeting. To obtain the face cut of the sheeting, set the compass at *j* and strike the part-circle, *g-h*. Parallel with *b-g* strike *h-i* and draw *f-i* at a right angle with *b-g*. Now strike *j-i*—the bevel at *i* gives the face cut of the sheeting. These bevels also give the cuts for the purlin.

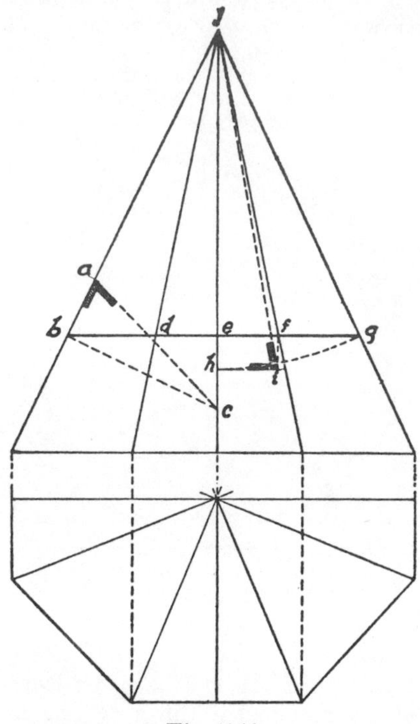

Fig. 141

Cuts for Purlin and Sheeting.—Fig. 142 shows one-half of a hexagon roof plan, bottom, and an elevation of the same roof at the top. To the left we have a purlin set in such a position that two of its edges are parallel with the rafter and two are at a right angle. To the right the purlin is set with its sides level and plumb. What we want to know is what bevels to use to make the different cuts. Let us take the problem to the left first. Strike *a-b* at a right angle with *A-C*. Using a convenient radius, strike the part-circle as shown. At a right angle with *A-B*, draw *c-i*, *b-h*, *x-e*, *j-g* and *d-f*. Parallel with *A-B* draw *h-i* and *f-g*. This done, draw *f-e* and *i-e*. The bevels at *e* give the two cuts. This being a half pitch roof, the bevels are the same, but if the roof were of a higher or of a lower pitch, the bevels would be different. It should be remembered that when the rafters join the purlin as they do in this case, against one side, then the bevels that are used to make the cuts for the purlin are the bevels that will also make the respective cuts of the roof sheeting.

To get the cuts for the purlin, shown to the right, is rather simple.

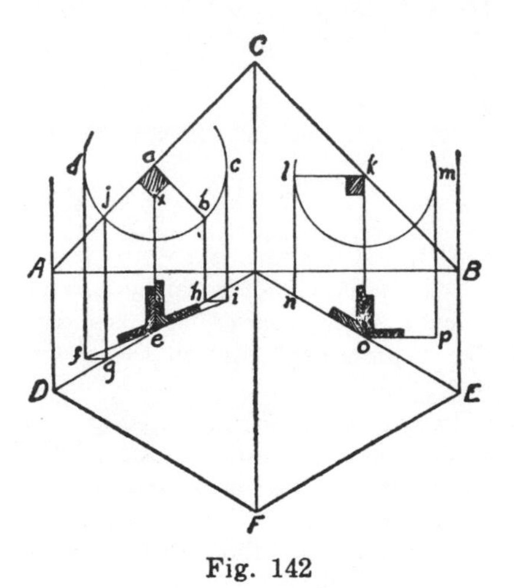

Fig. 142

Locate *k* about halfway between *C* and *B*, and with a convenient radius, strike the part-circle, *l-m*. At a right angle with *A-B* draw *m-p*, *k-o* and *l-n*. Draw *o-p* parallel with *A-B*. Now the bevel to the left of *o* gives the bevel cut of the purlin, while, as shown to the right of *o*, the other cut is a square cut.

Turret Roof.—Fig. 143 shows, *A*, an elevation and, *B*, a plan of a circular turret roof. The rafters in this roof are framed the same as for an octagon roof, with girts nailed between for nailing the sheeting to—

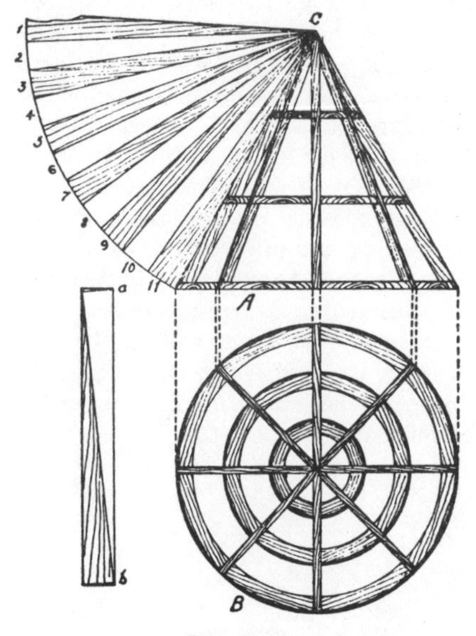

Fig. 143

for the sheeting of a cone-shaped roof must be put on up and down. A comparison of the plan and the elevation will reveal much of the construction. To the left of the elevation we are showing eleven sheeting boards spread out fanlike; every other one is shaded. The part-circle at the wide end is drawn with a radius equal to the length of the rafters. To the left, bottom, is a detail showing how to rip the boards. The dotted lines at *a* and *b* show the cuts of the wide ends of the boards made with a radius equal to the length of the rafters, giving the ends a circular cut. These boards must be slightly beveled on the edges if a good tight joint is desired.

LESSON 21

GOTHIC BARN ROOFS

Barn Roofs.—A treatment of the subject of roof framing could not be called complete without taking up the different kinds of barn roofs. While we have covered all of the roof framing principles necessary for framing any pitch roof that can be used on barns, we have not taken up the different kinds of roofs that are used on barns—for instance, Gothic, semicircular and Gambrel. These we will treat in their order before we leave the subject of roof framing. It will not be possible or necessary for us to take up the different kinds of constructions used for each of these types of barn roofs. Our purpose is to give a basis to work from, so that the roof framer by modifying can arrive at the construction that will best fill his particular needs.

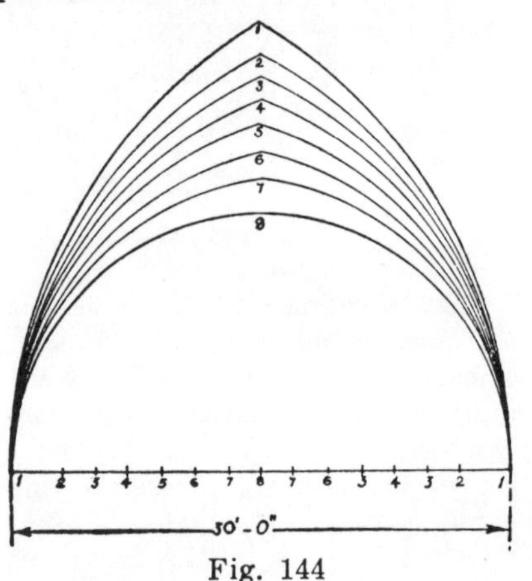

Fig. 144

Gothic Roof.—The Gothic roof, if well constructed, is a roof that gives a pleasing appearance. On the other hand, it is, perhaps, the most expensive roof construction that can be used. Especially is this true, in cases where the roof is to be self-supporting. It is even more expensive than a semi-circular roof. It does not give a great deal more over-head room than a gambrel roof; and a gambrel roof is more economical and perhaps more substantial. The plain gable

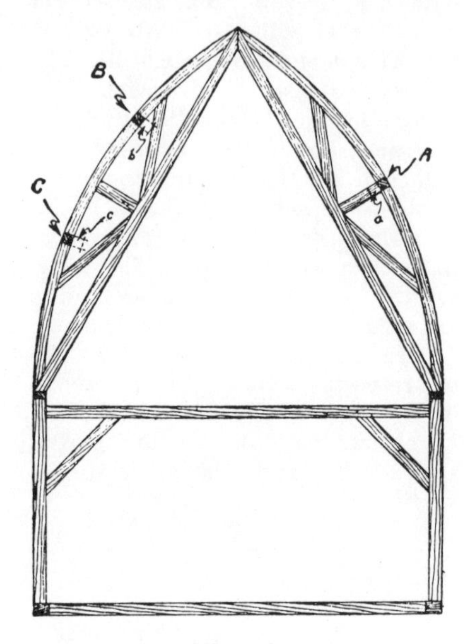

Fig. 145

roof, if the building is not too large, is the simplest roof construction for a barn, when service and economy play a prominent part in the choice of a roof.

Describing Gothic Roof.—Fig. 144 shows a simple way to obtain the radius for describing a Gothic roof. Presuming that we are framing the roof over a thirty-foot span; then, to obtain a full-pitch Gothic roof, let thirty feet be the radius. If you want a three-quarter pitch, use three-fourths of the span as your radius. A two-thirds pitch would, in this case, require a radius of twenty feet; while a half-pitch would completely lose the Gothic effect, for it would strike a semi-circular roof. The first of these pitches is obtained by pivoting the radius at number 1, to the right, and number 1, to the left. From these two points strike the part-circles; the second is obtained by pivoting the radius at 2, both to the right and to the left. In the same way use the right and left 3's, and so on to the last number, 8, which is the radius that gives the half-circle roof. This illustration simply shows that the roof framer can use almost any radius he chooses, that will satisfy his taste of appearance. By increasing the radius, the pitch can be increased beyond the full-pitch. The full-pitch Gothic roof, though, is perhaps used more than any other—so much so, that it can be said to be the standard pitch of Old Gothic.

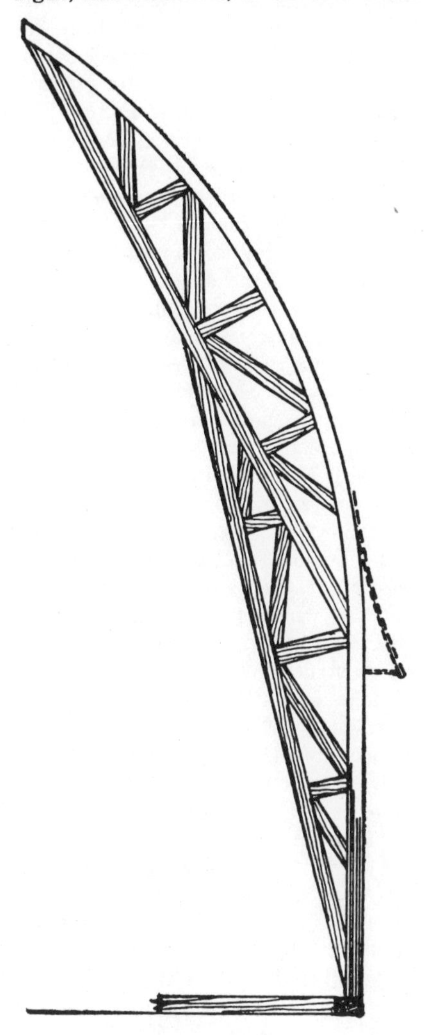

Fig. 146

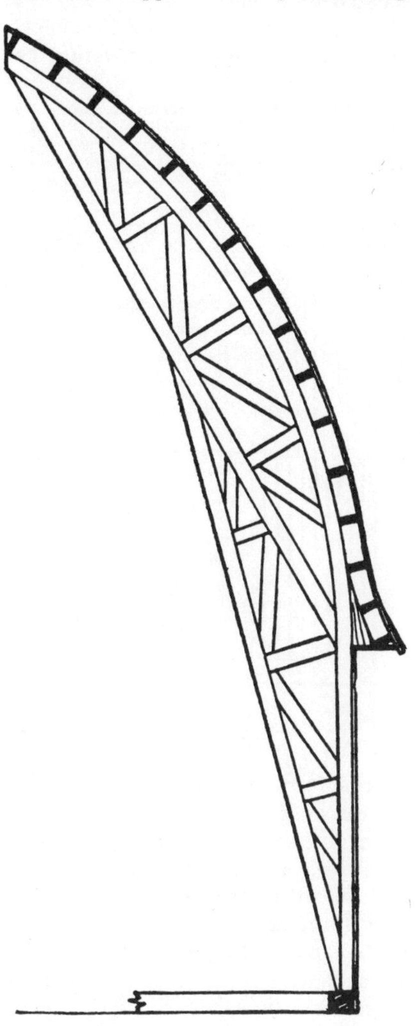

Fig. 147

Bent Construction.—Fig. 145 shows a construction of a section, or bent, of a barn with a Gothic roof. This is probably the simplest Gothic roof construction, and having a tie girder, it is not what is called a self-supporting roof. It will be noticed that the main rafters are exactly like the main rafters of a full-pitch gable roof. To the right, at *A*, we are showing how a girder, placed between the bents, will make it possible to use half-length rafters to fill in between the bents, and to the left, at *B* and *C*, we show how two girders will make it possible to use still shorter rafters. This is almost necessary in order to obtain the proper curvature of the rafters. By locating the girders as shown by dotted lines at *a*, *b* and *c*, the rafters can be nailed onto them, instead of butting against the girders, as shown by the first instances. Of course, this would necessitate extra supports for the girders.

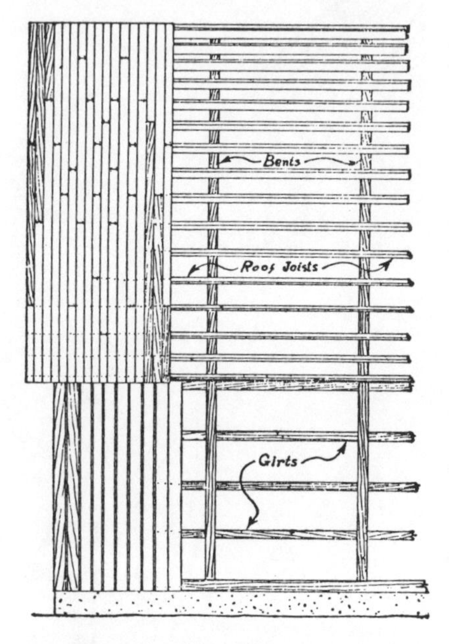

Fig. 148

Self-Supporting Gothic Roof.—Fig. 146 shows the construction of one half of a full-pitch self-supporting Gothic roof. We are not showing details of the joints, for that is a matter that must be determined by the material available, and the strength necessary to make the construction substantial. Eaves are suggested by dotted lines.

Supporting Roof Sheeting.—Fig. 147 shows another and better method of filling in between bents in order to support the roof sheeting. In this instance the sheeting is laid from eaves to the comb; whereas, in the case shown by Fig. 145, the sheeting would be laid horizontally. We have suggested an eaves construction, which is perhaps as simple as can be used. It should be remembered that the constructions shown by these illustrations are not hard-and-fast, but merely suggestive, leading the readers to plan constructions that will suit their individual needs; for no two readers will have exactly the same problems to solve.

A Side View.—Fig. 148 shows a side view, in part, of a barn partly boarded up, and the roof sheeting partly on. It also shows the sections, or bents, that support the roof joists. This illustration should be studied with the one shown by Fig. 147.

LESSON 22

SEMICIRCULAR ROOFS

The Coming Barn Roof.—The built-up, semicircular roof is perhaps the most economical roof construction for barns that can be used, excepting the plain double-pitch roof. In recent years this roof construction has been extensively used. Perhaps it could be said that this construction is fast coming into the lead for barn roofs.

Forming the Rafters.—Fig. 149 is a drawing showing the preparation that must be made for building semicircular rafters. The span, for convenience, is shown as 30 feet. The first operation is to strike a half-

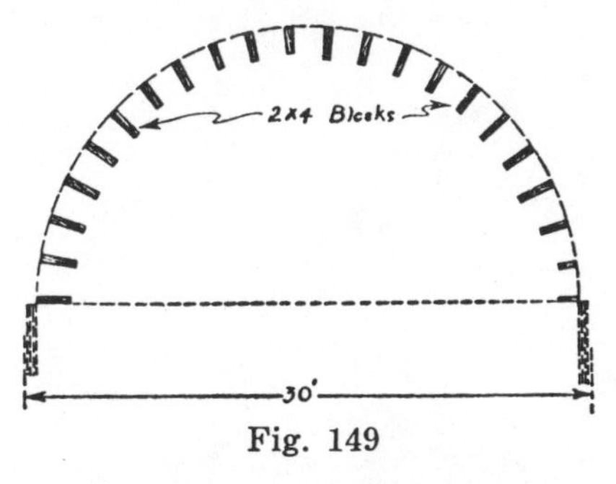

Fig. 149

circle, and then at regular intervals, fasten blocks, somewhat in the order shown on the drawing. When the barn is to have a loft, the floor of the barn loft can be used to work on, and the blocks can be nailed to it. When the blocks are in place, take good straight-grained ⅞-inch material, as long as can be obtained and as wide as the thickness of the rafters is to be. Spring these boards around the circle and fasten them to the blocks enough to hold their own. When the first ply of boards is on, start with the second ply of boards and nail it to the first one securely, being careful that no two end joints come closer together than 6 feet within any three consecutive plies of the rafter. The dotted lines to the right and to the left of the drawing, show how three of the boards lap down onto the posts of the barn. This

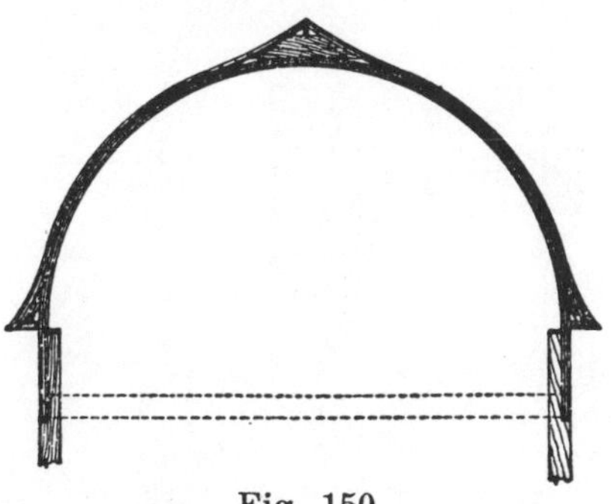
Fig. 150

is further illustrated by Fig. 150, where we show a completed semicircular rafter in place. The two dotted lines at the bottom represent the loft joists. A little study of this drawing will reveal the fact that this roof is self-supporting and free from any unnecessary obstructions.

Detail of Comb Construction.—Fig. 151 is a detail of the construction of the comb, if it can be called a comb. The saddle piece, marked *A*, should

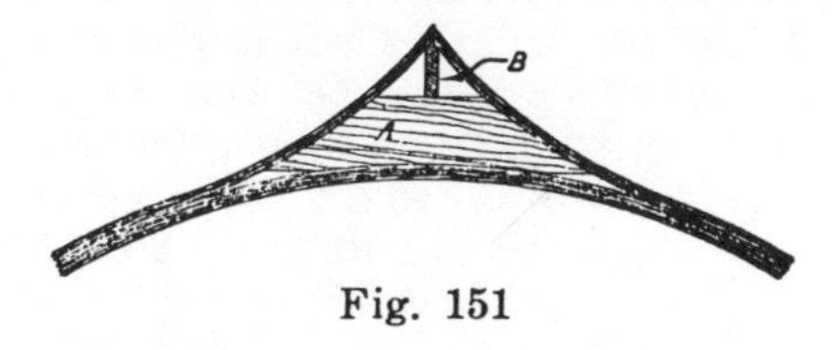

Fig. 151

be carefully cut out to fit the circle of the main roof and also the curve of the boards that form the point of the roof.

Other Details.—Fig. 152, to the right, shows a detail of the cornice.

It also shows how the three first boards extend down and are fastened to the studding, where the rafters are spaced the same as the studding. The roof sheeting is not shown in Fig. 151 and Fig. 152.

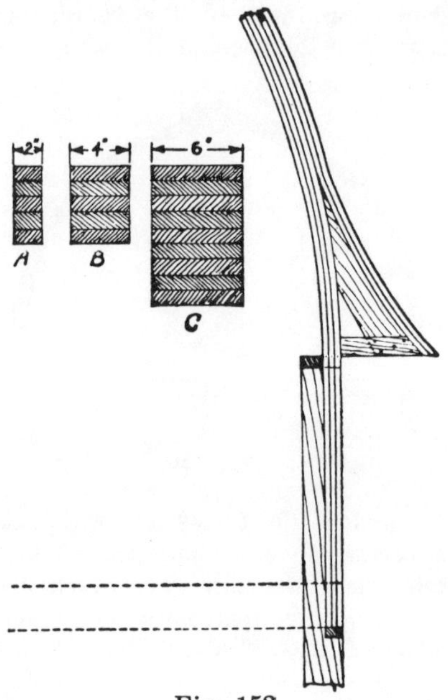

Fig. 152

At *A*, Fig. 152, we show a section of a 5-ply built-up rafter made of 1x2's. This construction is suitable for rather small buildings, when the rafters are spaced either 2 feet or 16 inches on center. At *B* is shown a section of a 5-ply semi-circular rafter built up of 1x4's. This construction of rafters, if spaced 2 feet on center, will support roofs with spans 30 feet and more. If, however, the width of the span is increased enough to justify adding an extra ply or two to the rafter, it should be done in order to insure a substantial roof construction. At *C* we are showing a section of a 9-ply semicircular rafter made of 1x6's. This construction is suitable for extra wide spans or in case the main rafters are spaced eight or ten feet on center and used as bents. In the latter case, the spaces between bents are filled in with roof joists and stripped, as we shall show in the next two illustrations.

Different Construction.— Fig. 153 shows, in part, a semicircular rafter for a bent of a large barn. At the bottom to the right, we show the cornice construction with four of the sheeting boards nailed on. The plancher is also in place. The roof joists are shown spaced 2 feet on center. The second ply of the rafter gives the rafter a sort of corbel, which helps to support the roof joists. This and the last ply are shown shaded. At *A-A* we are showing a section through *A—A*, showing to the right and to the left parts of the roof

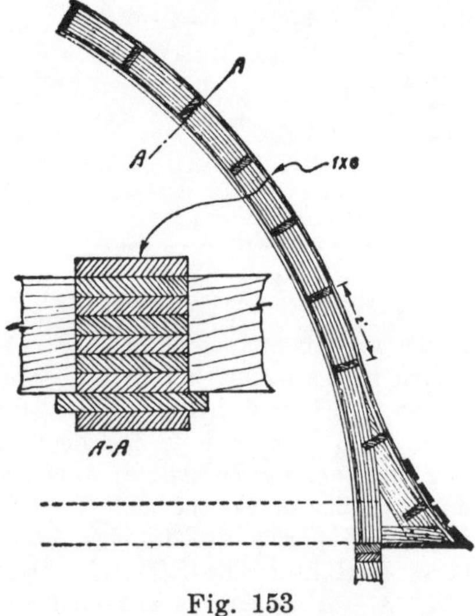

Fig. 153

joists. The dotted lines at the bottom of the main drawing represent loft joists.

Detail of Sheeting.—Fig. 154 shows the part of the roof shown by Fig. 153, looking from right to left. Here we are pointing out four roof joists toward the top, and at *A-A* we show where the bent is cut as shown in

Fig. 153, using the same reference letters. The 1x6 strips to which the roof sheeting is to be nailed, are

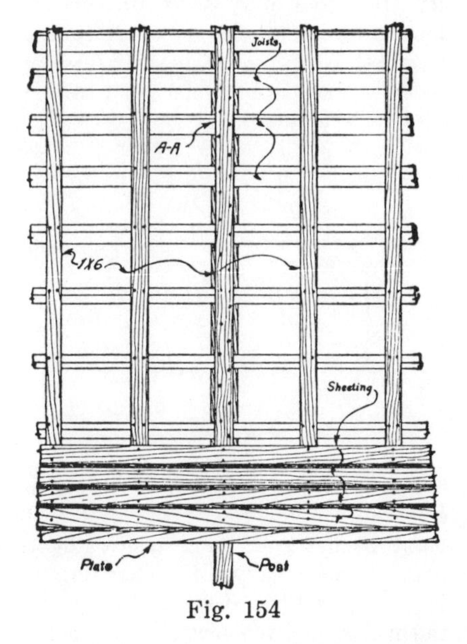

Fig. 154

pointed out at the center. Four sheeting boards already nailed on are shown toward the bottom, also the plate and the post supporting the rafter for the bent.

Governed by Conditions. — There isn't anything hard-and-fast about the constructions we have shown, and that is as it should be, for we are dealing with principles pertaining to semicircular roofs, and not with the needs of those who might want to use such roofs on barns or other buildings. It is a self-evident fact to carpenters, and should be to most laymen, that a short span requires less strength to support the roof than does a long span. And just as the spans, in actual practice, necessarily must vary, just so the strength of the rafters must be increased or decreased to suit the different situations. Wind pressures, storms, snows, earthquakes or their absence must be taken into consideration in designing semicircular rafters—also the strength or weaknesses of the materials to be used, all of which are factors in constructing semicircular roofs, or any other kinds of roofs, for that matter.

LESSON 23

GAMBREL ROOF DESIGNS

Suitable for Barns.—In many ways the gambrel roof is well suited for large barns, because it can be well braced without obstructing the space through the center, which is usually used for carrying hay back into the loft by means of a hay fork hooked to a carrier that runs on a track installed just below the comb of the roof.

Designs Unlimited.—It is claimed by some builders that a gambrel roof

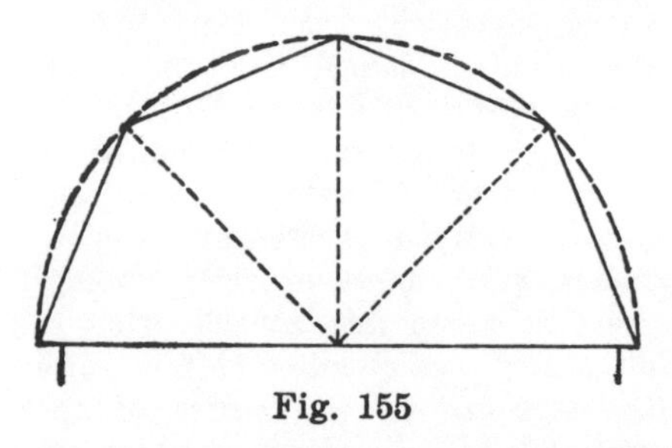

Fig. 155

can not be built so it will have a pleasing appearance. Of course, that is a matter of taste. We believe that the reason why so many gambrel roofs are lacking in good appearance is because the designers limit them-

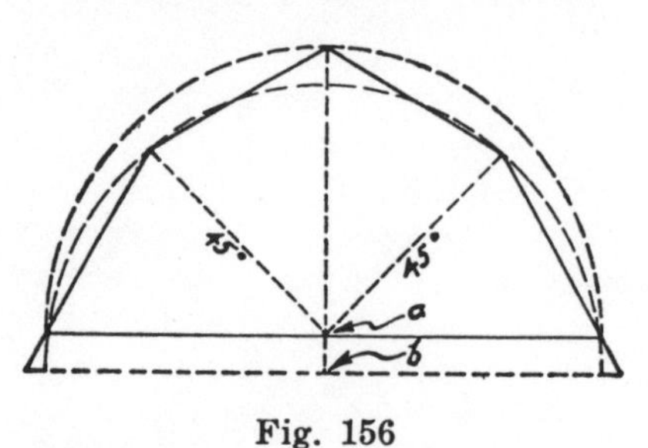

Fig. 156

selves to just a single design, or at best, to not more than just a few. The field for gambrel roof designs is large and the resourceful designer can easily adjust his designs to suit almost any condition that might exist. It is our purpose to present solutions to the problems of designing gambrel roofs, so as to bring about more nearly acceptable designs for this style of roofs.

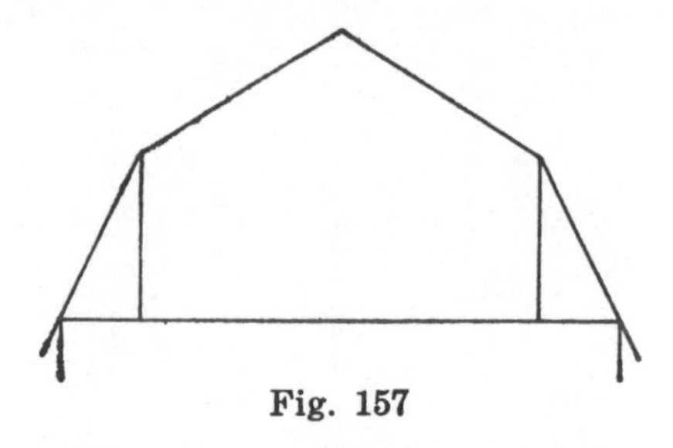

Fig. 157

Poor Design.—We have met builders who contend that a gambrel roof must be designed on a true octagon pattern, such as we are showing by Fig. 155. A little study of this drawing will reveal the fact that the steep

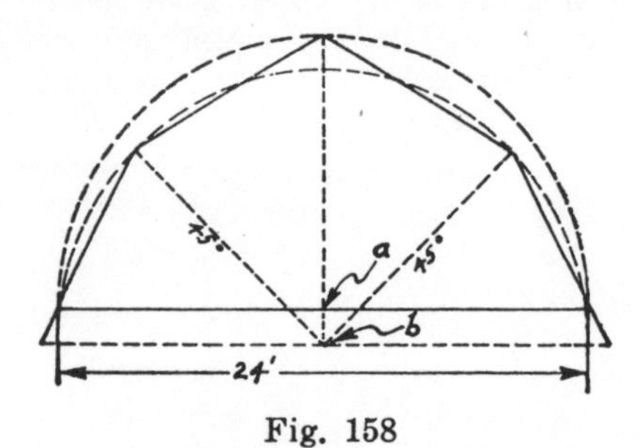

Fig. 158

sides of this design are too steep, while the flat sides are too flat both for producing a pleasing appearance and for throwing off water.

Better Design.—Fig. 156 gives a different method for laying out a gambrel roof design. Here two half circles are struck, one from point *a* and the other from point *b*. Three lines are struck from point *b*, a perpendicular line and two lines on a 45-degree angle as shown. With these lines in, draw the continuous lines that give the design of the roof. It should be noted here, that point *a* is in line with the top of the plates, while point *b* is in line with the bot-

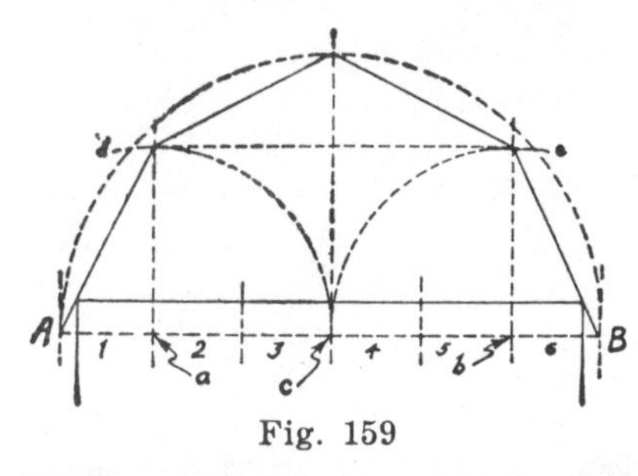

Fig. 159

tom edge of the eaves. This design is shown, stripped of all unnecessary lines, by Fig. 157. Comparing this with Fig. 155, these improvements can be observed: The steep pitch has been reduced and the flat pitch has been increased, while the sides are all the same in length.

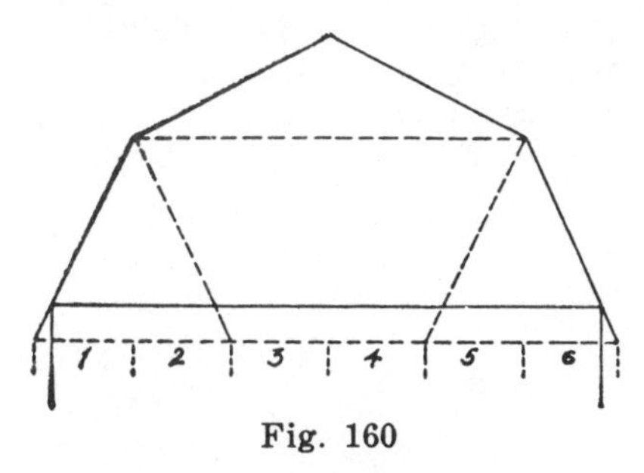

Fig. 160

A Little Different Design. — Fig. 158 has been designed practically on the same basis as Figs. 156 and 157, excepting that the perpendicular and two 45-degree lines were struck from point *a,* instead of point *b.* This change improves the design, making the steep sides longer and the flat sides shorter.

Another Designing Method.—Fig. 159 introduces another method of designing gambrel roofs. Here the distance between the bottom edges of the eaves, *A* and *B,* has been divided into six equal parts. This done, strike the half circle, *A-B,* from point *c*—then from *a, b* and *c* raise three perpendicular lines, as shown. From points *a* and *b,* strike *c-d* and *c-e.* Now lay off the roof design, by drawing the continuous lines, to the points shown. This design gives you a full pitch for the steep sides and a one-

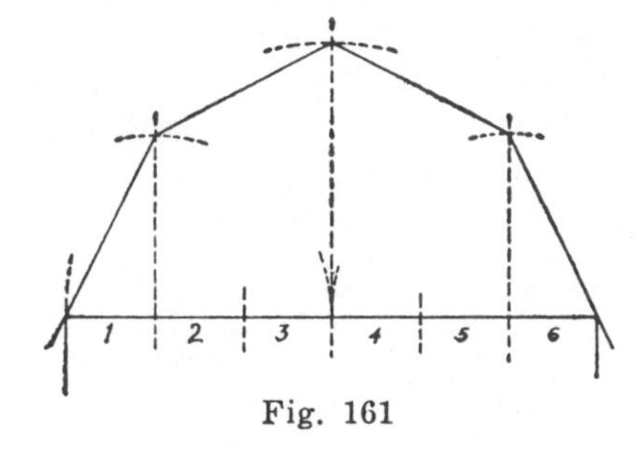

Fig. 161

fourth pitch for the flat sides, as shown by Fig. 160.

Different Application. — Fig. 161 shows the same method that was used in Fig. 159, but applied somewhat differently. Here the span has been divided into six equal spaces, whereas

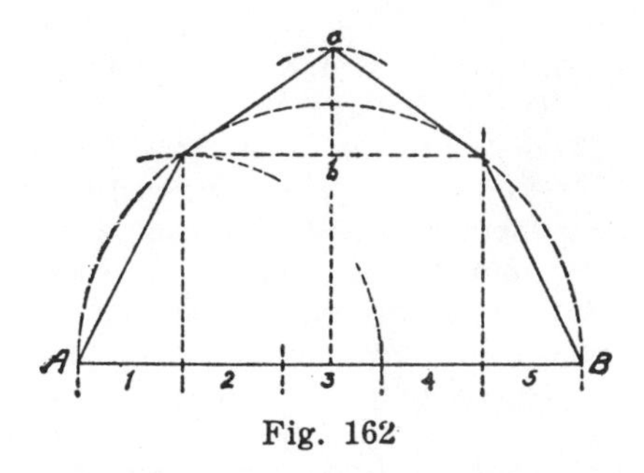

Fig. 162

in the other figure the distance between the bottom edges of the eaves was divided into six equal spaces. The points for laying off the design are obtained just as explained in Fig. 159. The steep sides are still full pitch, but a little longer than in the other case, and the flat sides are still one-fourth pitch, but a little shorter, which improves the design.

Method Modified.—Fig. 162 shows the use of the same method with the span, *A-B*, divided into 5 equal spaces. The run of the steep side is one space and the rise is two spaces, just the same as shown in Figs. 159 and 161. But the run of the flat sides is only one and one-half spaces and the rise is one space. This gives the steep sides a full pitch and the flat sides a one-third pitch. The distance between *a* and *b* is made equal to one space.

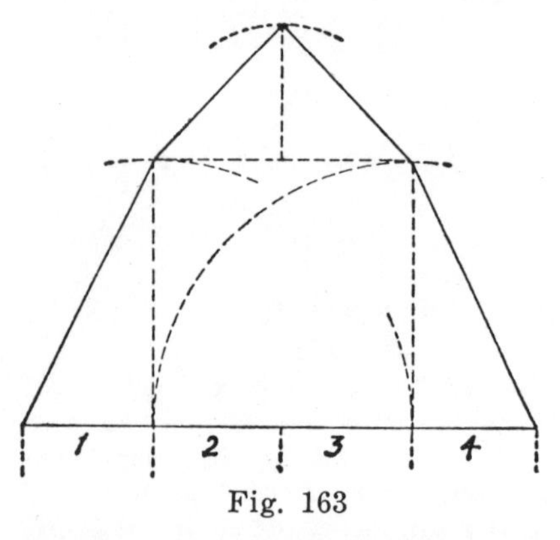

Fig. 163

This design is more suitable for throwing off water than any we have shown previously, and the appearance is also satisfactory.

Design with Increased Pitch.—Fig. 163 shows the method carried one step farther, where the span is divided into four equal parts. The steep sides have a one-space run and a two-space rise, while the flat sides have a one-space run and a one-space rise, making a one-half pitch. This design, while not so satisfactory from the standpoint of appearance, nevertheless, has its advantages under certain conditions.

Knuckle or Curb Joints.—Fig. 164 shows two good knuckle joints for gambrel roofs, when a purlin is used. They are both substantial, but the one to the right is perhaps the better joint.

Fig. 165 shows two good curb joints for light gambrel roofs when no purlin is used. The joints are

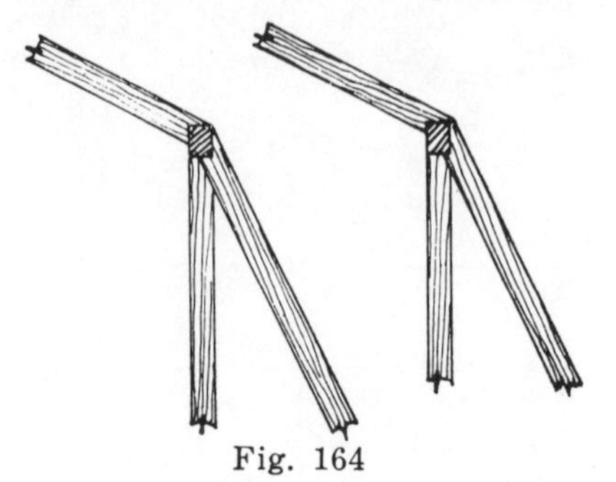

Fig. 164

shown naked; in practice they should be reinforced with cleats on one or on both sides, as conditions might require.

The top and the bottom cuts of the steep rafter shown at *A*, are obtained by using the figures on the square corresponding with the pitch of the steep sides, and the cuts for the rafters of the flat sides are obtained by using the figures giving the flat pitch. The cut pointed out at *c* is the same as the cuts of the steep pitch, but the

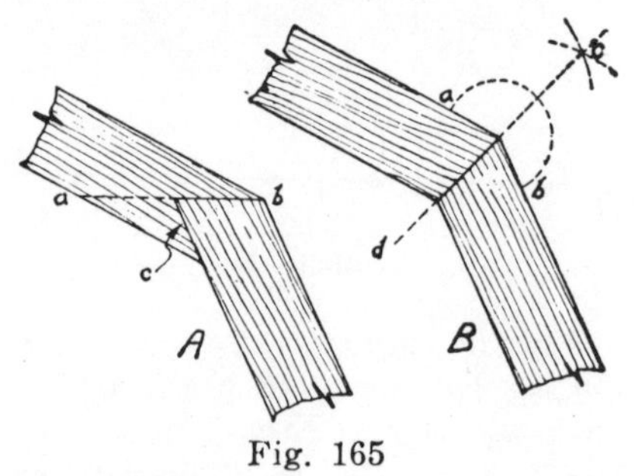

Fig. 165

square must be applied to line *a-b*.

The cut for the joint at *B* is what is called a member cut. To obtain this cut, strike the part-circle, *a-b*, and with a compass set at points *a* and *b*, respectively, establish point *c*. Now draw line *c-d*, and you have the member cut that will make the joint.

LESSON 24

FLAT ROOF CONSTRUCTIONS

Flat Roof Work.—Flat roof framing is something that every carpenter must do sooner or later in his career. This branch of roof framing is no less important than pitch roof framing, for insofar as the relative amount of carpenter work involved is concerned, it is a close second. Most business houses, large and small, have

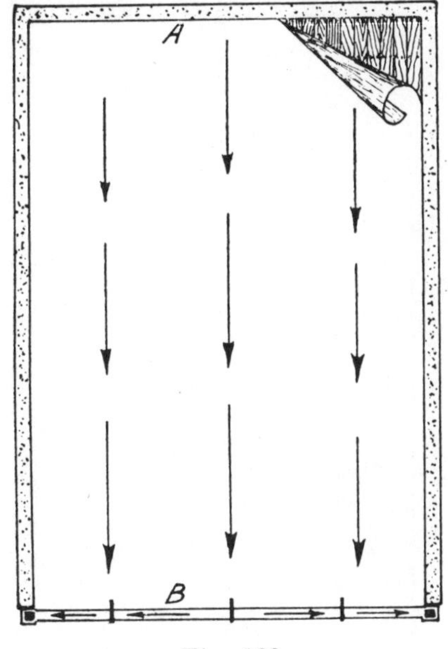

Fig. 166

some sort of flat roof. The same is true of many factory buildings, while practically all shop buildings have flat roofs.

Plan of Flat Roof.—Fig. 166 is a simple plan of a small flat roof. The high point is indicated at *A* and the low point at *B*, which necessarily would make the roof fall in the direction of the arrows. To the upper right we show a roll-back of the roofing, which reveals the decking of the roof. The gutter is shown at the bottom, where a two-way fall is indicated by the arrows.

Trussing.—One method of trussing a roof like the one shown by Fig. 166 is shown in Fig. 167. Here *A* again

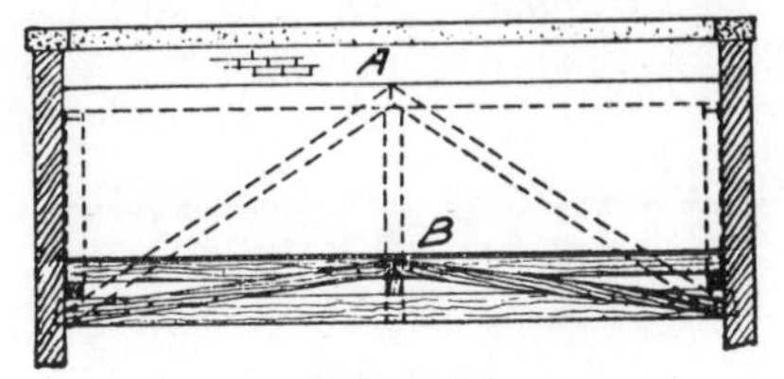

Fig. 167

represents the high point and *B* the low point. The dotted lines show the construction of the highest truss, while the lowest truss is shown by continuous lines and shaded. In this construction each roof joist is trussed as shown, which means that each truss from *A* down to *B* must be just a little lower than the truss that comes before it. In this construction the braces are in compression, but in the truss construction shown in Fig. 168 the braces are in tension, which is the only difference between the

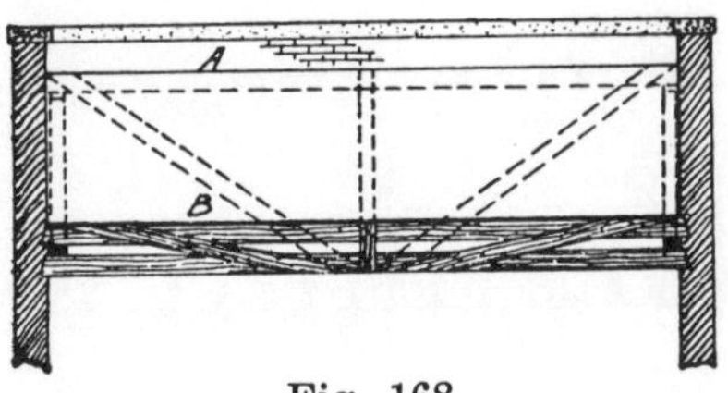

Fig. 168

two constructions, excepting that the roof joists shown in Fig. 167 are not supported by the wall, while in Fig. 168 they are shown supported by the wall just as the ceiling joists are. The dotted lines at the right and left show how the ceiling joists can be supported by means of shores. Either of these two constructions are good, but where the roof joists are supported at the ends by the wall they should be anchored to the wall, which adds some bracing to the firewall, should it be needed.

Weak Point.—Fig. 169 shows a longitudinal section of the roof plan

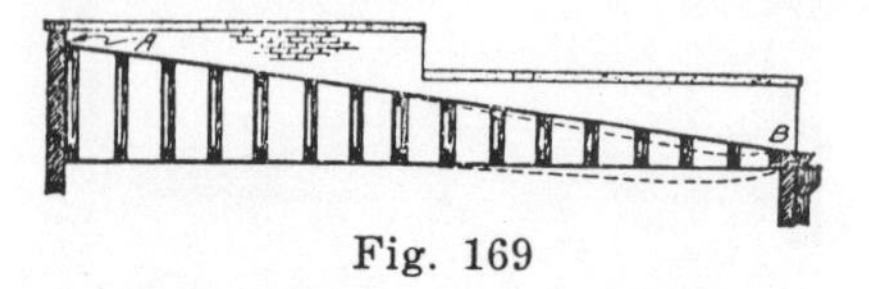

Fig. 169

shown in Fig. 166 and the truss construction shown in Fig. 167. The slightly-curved dotted lines shown to the right indicate where this roof construction is weak. This is true of both truss constructions just explained. The closer the truss is to the low point, *B*, the weaker it becomes, because the bracing value is reduced just as the rise of the trusses is reduced. On the other hand, the closer the truss is to the high point, *A*, the stronger the truss becomes, because the bracing value increases just as the rise of the trusses increases.

There are several ways to reinforce

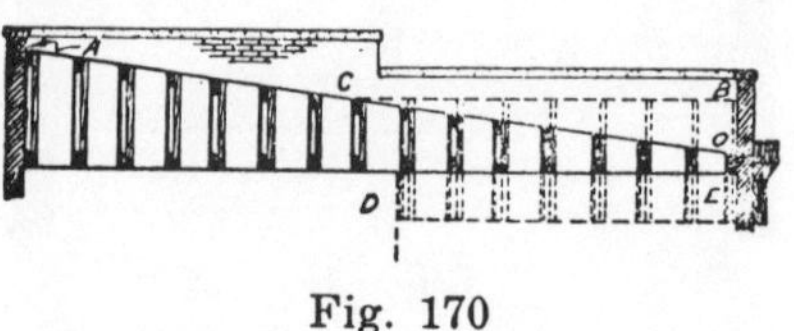

Fig. 170

flat roofs toward the low point, where in time a sag will form and pool the water. When the floor space is divided into rooms, usually at the rear of the building, a supporting partition or two running at a right-angle to the trusses will support this part of the roof and prevent sagging. But where the floor space is not divided into small rooms at the rear, some other means must be used to reinforce the weak part of this roof.

Reinforcing Flat Roofs.—Fig. 170 gives two ways for reinforcing flat roofs. From point *A* to point *C*, the trusses are strong enough to prevent sagging of the roof, but from *C* to *O* they are not. When the room is divided by a partition at about point *D*,

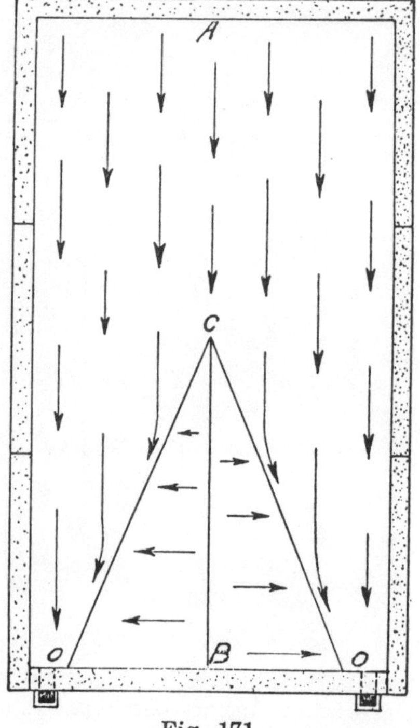

Fig. 171

then the remedy would be to lower the ceiling joists from *D* to *E*, as indicated by dotted lines, which would increase the rise of the trusses and relatively their strength. When a dropped ceiling is impracticable, then the trusses from *C* to *B* should all be made with the same rise. This would give the roof a plan like the one shown by Fig. 171. Compare Figs. 170 and 171 where the reference letters indicate the same points in both figures. The arrows in Fig. 171 show the fall of the roof.

Truss Construction. — Fig. 172 shows the construction of the roof shown in plan in Fig. 171. Here the

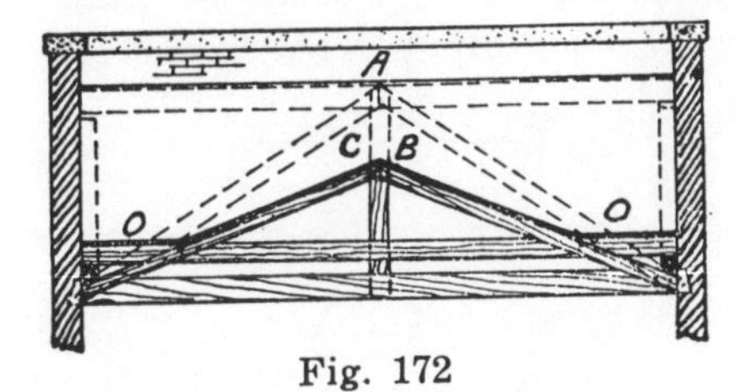

Fig. 172

rise of the trusses is gradually reduced from *A* to *C*, and from *C* to *B* it is kept the same. Points *O*, *O*, in both Fig. 171 and Fig. 172, indicate the outlet for the water through the firewall. This construction will pre-

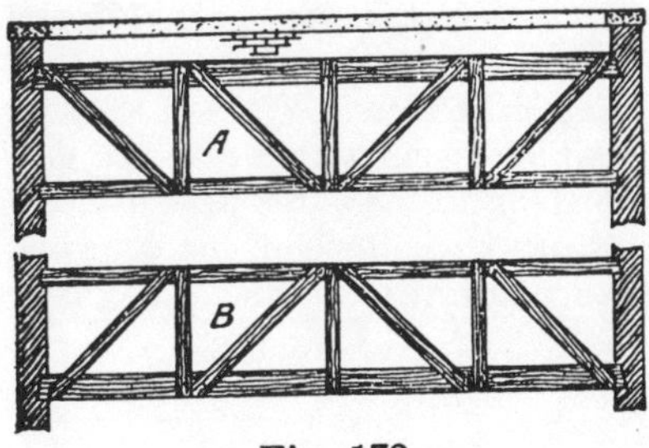

Fig. 173

vent sagging at the low point of a flat roof, and is the best way that I know of, for reinforcing such roofs.

Other Constructions. — Fig. 173 shows two methods of trussing for flat roofs. At *A* the braces are in tension, while at *B* they are in compression. In both of these instances the ends of both the roof joists and the ceiling joists are supported by the wall, which should be anchored to the wall at regular intervals.

Fig. 174, *A*, shows a truss that is a combination of the two trusses shown in Fig. 173. Half of the braces are in tension, while the other half are

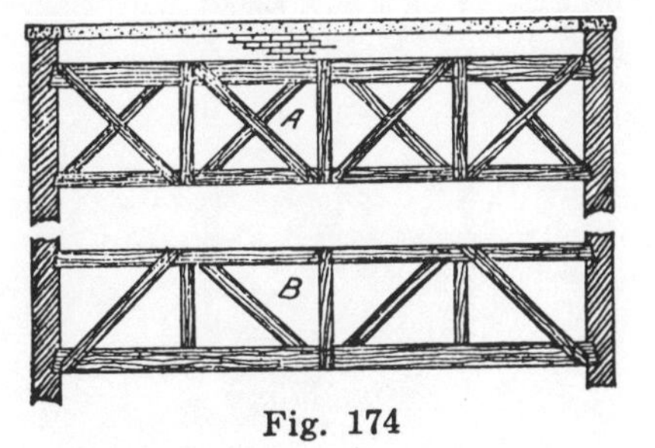

Fig. 174

in compression. The truss shown at *B* is also a combination of the two trusses shown in Fig. 173, but of a lighter construction.

Methods of Reinforcing. — There are two ways that sometimes are used for reinforcing flat roofs to prevent sagging, but we do not like either of them. One is to use a beam at the center of the last six or eight roof trusses, supported at one end by the rear wall and on the other end by a column. The other is to build a truss above the roof and hang, as it were, the roof to it at the center between the walls.

LESSON 25

FLAT ROOFS OVER REINFORCED CONCRETE

In the previous lesson we were dealing with flat roofs supported by wooden trusses, and the principal emphasis was on how to build the trusses so as to make them hold the roof in its proper position, but in this lesson we are taking up flat roofs that are supported by reinforced concrete slabs.

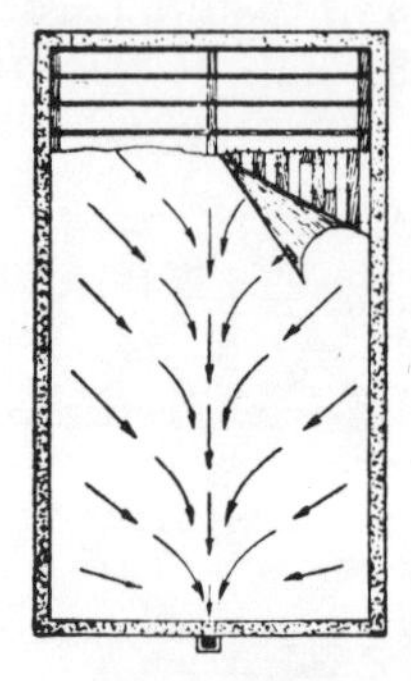

Fig. 175

Plan of Roof.—Fig. **175** shows a plan of a roof with one outlet at the center of the rear. At the upper part of the drawing we are showing the roof joists resting on their supports. To the right we show a roll-back of the roofing, which reveals the sheeting and roof joists. The arrows indicate the flow of the water toward the outlet at the rear, and consequently the slopes of the roof. Fig. 176 shows a cross section of this roof at its lowest point. The arrows indicate the flow of the water. The horizontal dotted line gives the bottom edge of the highest roof joists, while the perpendicular dotted line at the center gives the line of the gutter of the roof to the outlet. Compare Figs. **175** and 176.

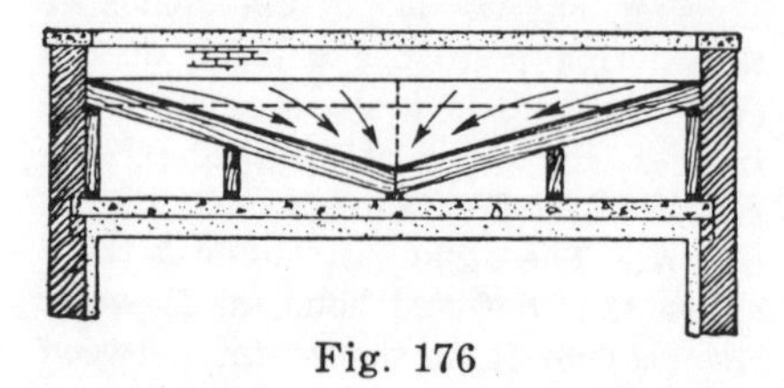

Fig. 176

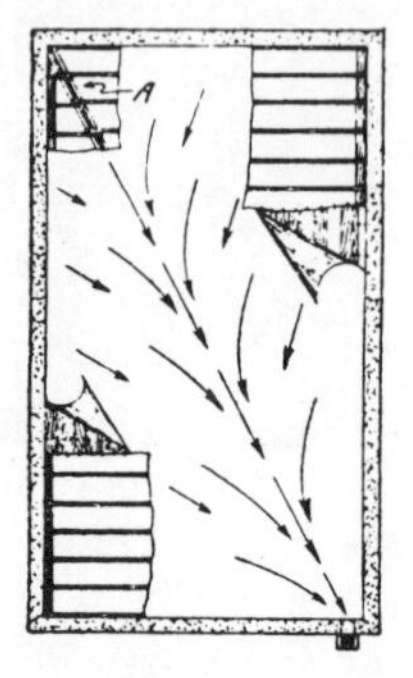

Fig. 177

Diagonal Gutter.—Fig. **177** shows a similar roof plan with a diagonal gutter leading to the outlet shown to the right at the bottom of the drawing. The support for the diagonal gutter of this roof is pointed out at *A*. The roofing has been cut out at three different places to show the construction. The sheeting can be seen at the two places where the roofing is rolled back. A cross section of this roof at the rear of the building is shown by Fig. 178, where the diagonal gutter

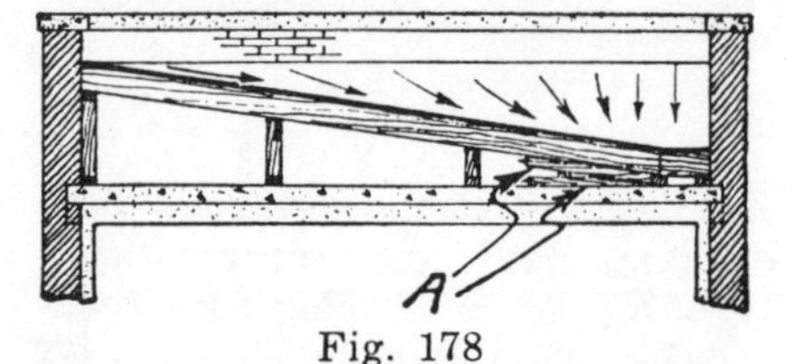

Fig. 178

support is pointed out in part, at *A*. Figs. 177 and 178 should be compared and studied.

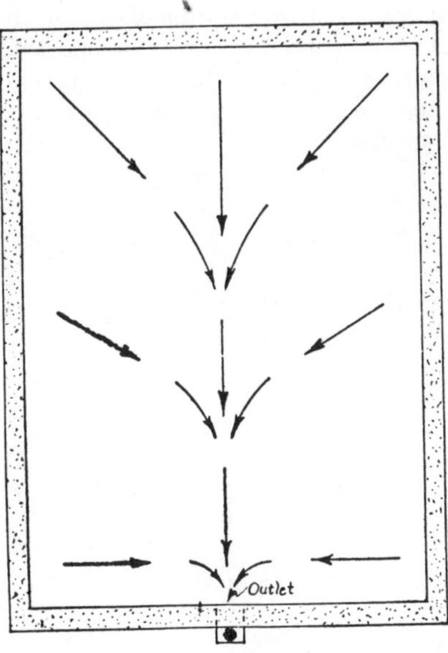

Fig. 179

Center Gutter.—Fig. 179 shows a plan of a roof which slopes somewhat as indicated by the arrows. This roof is more nearly on the order of a huge gutter. It slopes to the center, and

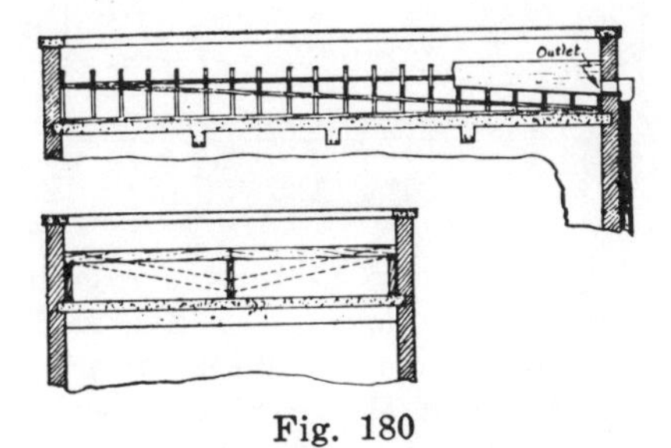

Fig. 180

the center inclines to the low point, carrying the water to an outlet through the fire wall, as shown on the drawing. Fig. 180 gives two sections of this roof. At the top the roof is shown cut at the center. To the right is shown a part of the sheeting, covering six joists. The rest of the joists to the left are uncovered, showing how the inclination is gradually

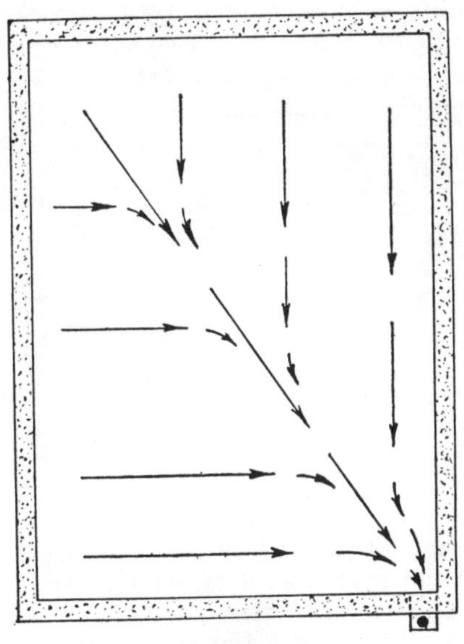

Fig. 181

reduced until the high point is reached at the extreme left, where the joists are on a level.

The bottom drawing shows a cross section with the side and center sup-

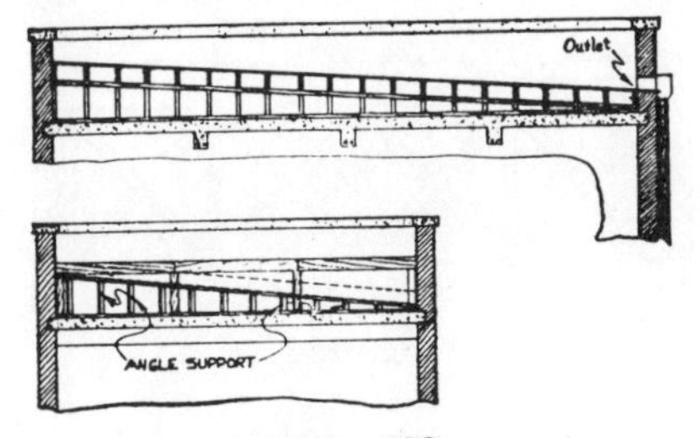

Fig. 182

ports in place and the two highest joists set. The inclining dotted lines show the position of the joists at the lowest point of the roof. It will be noticed by studying the two sections, that the side supports are the same in height the full length of the roof, while the center support inclines gradually toward the low point where the water escapes through the outlet. This construction is suitable for buildings that have a good base to work on, especially through the center toward which the water drains.

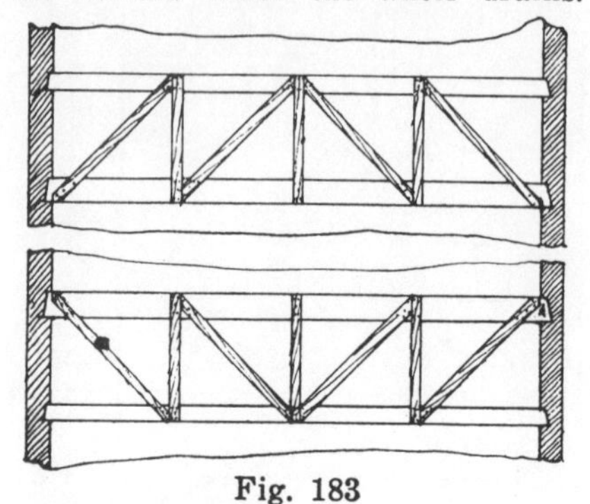

Fig. 183

Both the outlet and the downspout should have a good capacity so as to dispose of the water quickly.

A different flat roof construction is shown by Fig. 181. Here the gutter runs in a diagonal direction, as indicated by the arrows. The outlet is shown to the right at the low point. Fig. 182 shows two sections of this roof construction. The upper one shows how the right half of the roof inclines toward the outlet, while the bottom one shows the joists in place at the high point. By dotted lines the inclination of the joists at the low point is indicated. The inclined angle support is pointed out by arrows.

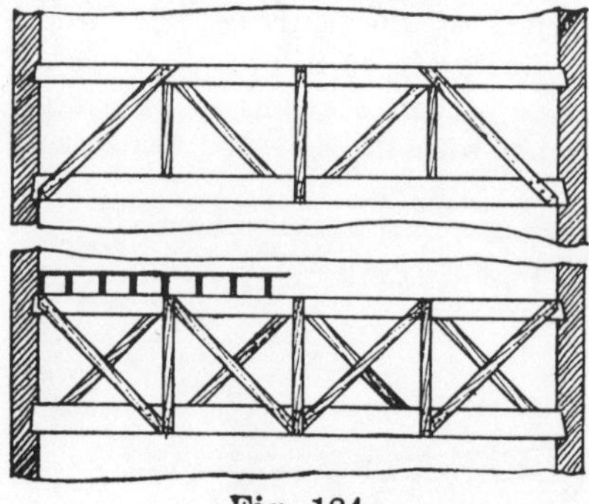

Fig. 184

Trussing Flat Roofs.—Two flat roof trusses are shown by Fig. 183. The top one shows the braces placed in such a way that their supporting qualities depend on compression, while the bottom one shows the braces in tension. The upright ties act under compression where the braces are in tension, and vice versa. Both of these designs are good; the upper one, however, is probably the most commonly used. Modifications of these trusses are both possible and practical.

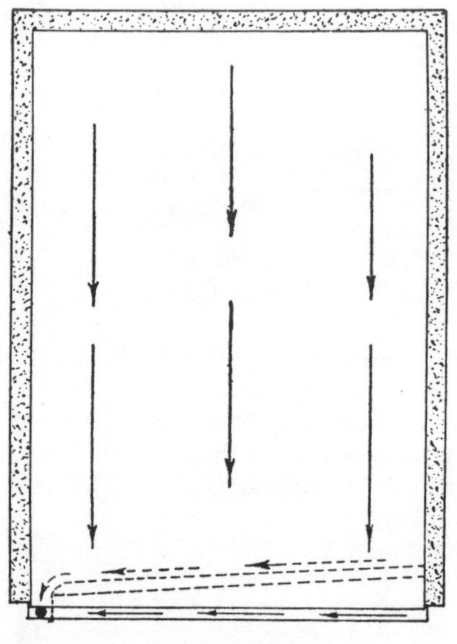

Fig. 185

Fig. 184 shows at the top a good truss construction for a flat roof. Here two braces act under compression, and two are in tension, while two ties are in tension and one is under compression. At the bottom is a modification of the truss just explained, in which the bracing is doubled. This makes a good truss, and is suitable for carrying sections of joists, as shown to the upper left.

Flat Roof and Hanging Gutter.—

Fig. 185 shows a plan of a flat roof. The arrows show the slope of the roof. At the bottom of the drawing is shown a hanging gutter, with arrows indicating that it slopes to the left. The dotted lines, almost at the bottom, show how a planted on gutter can be used on this roof. The dotted-line arrows show the gutter's fall.

Simple Trussing.—Fig. 186 shows a commonly used flat roof construction. At the top is shown the highest roof joist, trussed and tied to the ceiling joist directly below it. At the bottom is shown the lowest roof joist, trussed and tied to the ceiling joist below. The truss shown at the top will carry a load much greater than will ever come on it under ordinary conditions, but the bottom truss is not strong enough to support its load without reinforcement. In fact, the strongest truss of such a roof carries the lighter load, and the weakest truss carries the heavier load; that is, assuming that every roof joist is trussed in the same way, which is usually the case. Sometimes the lower part of a flat roof is reinforced by supporting partitions, which eliminates the danger of sagging. However, when there are no partitions under this part of a flat roof, the trusses should be reinforced, either with extra bracing, or by putting the trusses at closer intervals. It is not an uncommon thing to find a flat roof so badly sagged at the lower part that it will not drain, consequently water pools, which soon develops leaks in the roof, causing much damage to the ceiling or even to the goods stored in the building. Reinforcing beams placed under this part of the roof will remedy such a condition, but it is much better to make the original construction substantial enough to carry the load. Perhaps the best way to prevent sagging at the lower part of a flat roof is to increase the distance between the ceiling joists and the roof joists. This can be done by putting the roof joists farther above the ceiling joists, or by lowering the ceiling joists from the point where the trusses begin to show weakness. The objection to this, though, is that it would bring an offset in the ceiling line of the room where the ceiling joists are lowered. In most cases the back part of such rooms are partitioned off from the front, as in store buildings, and then the offset in the ceiling will not be noticeable.

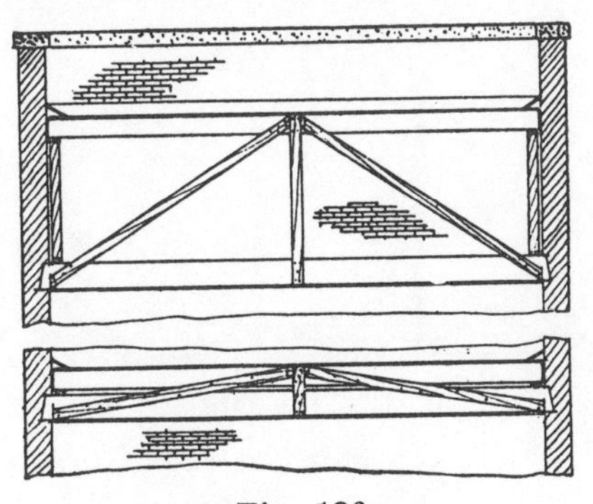

Fig. 186

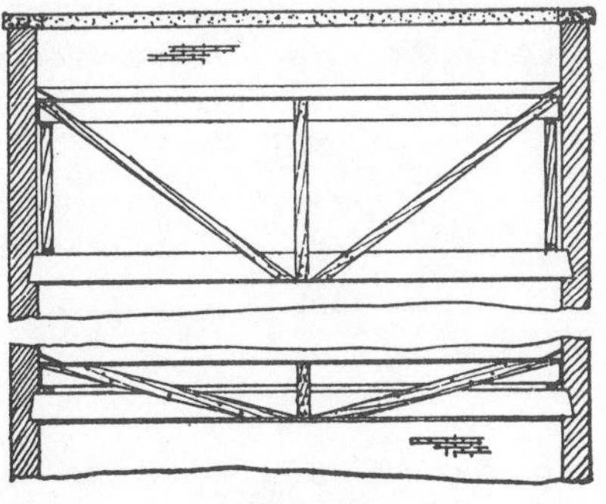

Fig. 187

Fig. 187 shows an inverted truss for supporting flat roofs. As a rule, there is little difference between the strength of this truss and the one shown in Fig. 186, so long as they both depend entirely on nails for their holding qualities.

LESSON 26

SEGMENTAL LATTICE TRUSSES

Wood and Steel Trusses.—As carpenters we are interested in our own trade and in holding for our own craftsmen as much of the work as possible. In recent years, much of the work that used to belong to the carpenters has either been eliminated or shifted into the jurisdiction of some other trade. This is especially

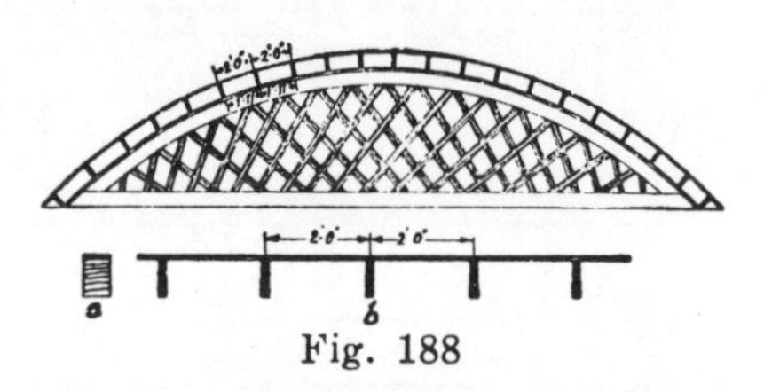

Fig. 188

apparent in public buildings and business houses where cement and steel are, to a great extent, taking the place of wood. Speaking of trusses, steel has taken over the greater part

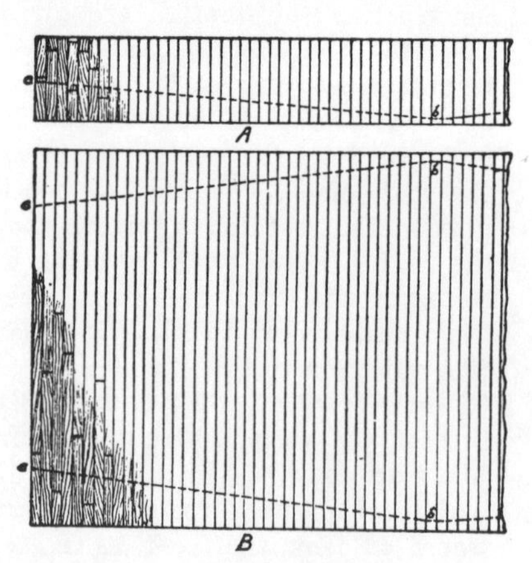

Fig. 189

of that field. Notwithstanding that, wood is still being used for trusses in many instances. During World War II there was a periodic revival of wooden trusses, which indicates that wooden trusses do give substantial service. In fact, there are many cases in which wooden trusses are much more economical than steel trusses, and just as serviceable, if not more so.

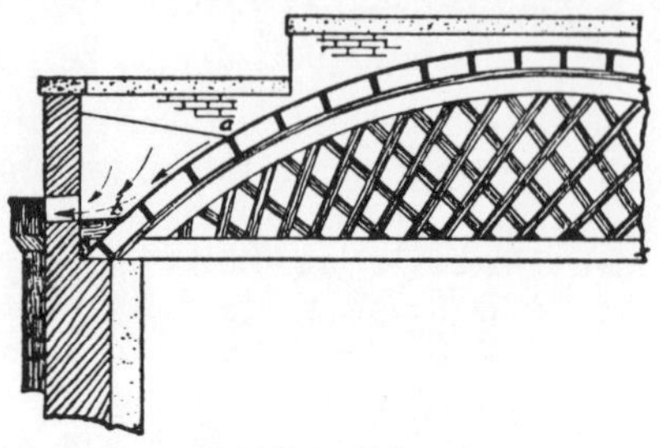

Fig. 190

Segment Truss.—Fig. 188, main drawing, shows a lattice segmental truss. To the upper left, we show two spaces with the width of the spaces shown in figures at the sheeting line. At the bearings the same two spaces are shown, where they are

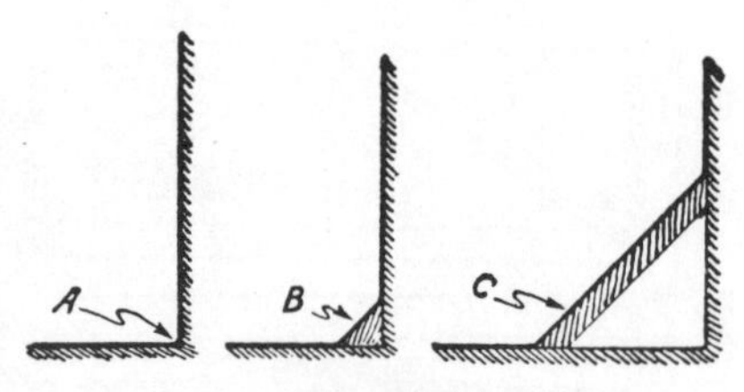

Fig. 191

only 1 foot 11 inches, instead of 2 feet from center to center. The reason for this is that the joists radiate from a common center. If the joists are spaced 2 feet from center to center on the plate that is nailed over the truss, on which the joists rest, then the joists will be spaced more than 2 feet on the sheeting line, and for that reason there will be a great deal of waste in lumber when the roof is sheeted.

Spacing Roof Joists.—How to determine the distance of the spaces on the joist-bearing plate over the truss, so that the upper edges of the joists will run 2 feet from center to center is shown by the two details of a templet at *a* and *b*. At *a* is shown an end view of the part shown at *b*. The templet is made of a 1x6 with blocks of the joist material nailed to it 2 feet on centers. When the templet is completed, spring it over the truss in such a manner that the blocks will contact the plate and fasten it. Now measure the distances from center to center of the blocks where they contact the plate and take the average distance of these spaces for spacing the joists on the truss. When the

Fig. 192

joists are set you will find that the sheeting will go on without much waste.

Side View.—Fig. 189 shows a side view, *A*, of a segment roof, in part, with the sheeting on, supported by trusses similar to the one shown in Fig. 188, and *B* shows a plan, in part, of this roof. The dotted lines show where the gutters come so as to carry the water to the outlets marked *b*. The *b*'s represent the low points of the gutters, while the *a*'s give the high points. Fig. 190 gives a section of such a gutter through the low point. The arrows show the fall of the gutter, while *a* and *b* give the high and low points. Compare this figure with one of the gutters shown by dotted lines at *B*, Fig. 189—the letters refer to the same points in both instances.

Cant Board.—In all of the draw-

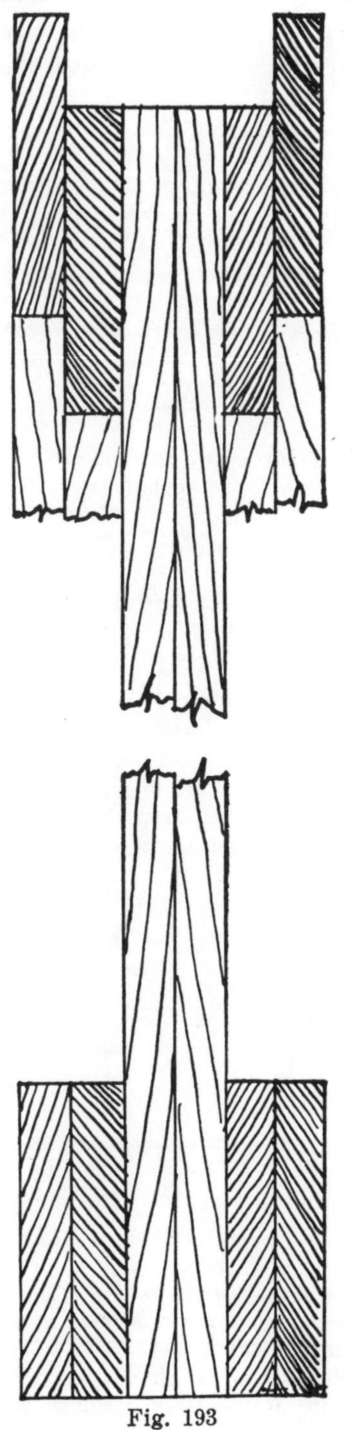

Fig. 193

ings of this lesson and the one preceding this one, we omitted showing

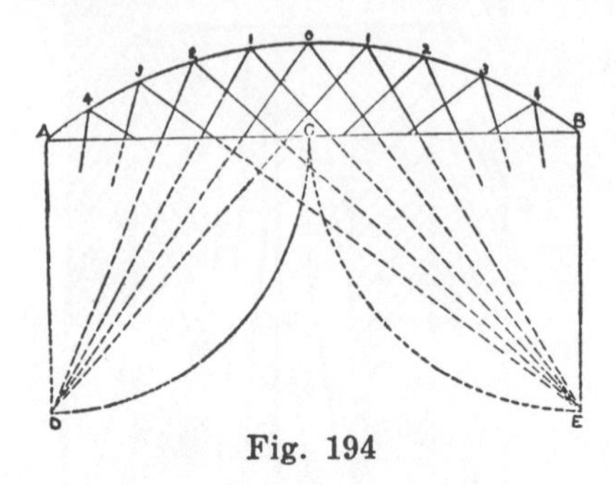

Fig. 194

the cant board or cant strip, which is used in the angle between the roof

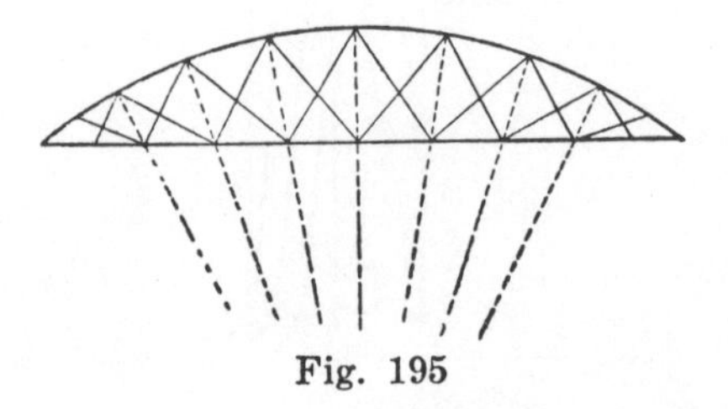
Fig. 195

sheeting and the firewall. There are three ways of preparing the angles of a flat roof for the roofing, which we are showing by Fig. 191. At *A* is shown an angle finished at a right angle. This is not a bad method, if the roofing is placed carefully and there are no large openings between the sheeting and the masonry; however, the sheeting should not hug the masonry. The danger in this construction is that the roofing might break at the angle. This is especially true if roll roofing is used and put on during cold weather. In fact, roll roofing should not be put on excepting in reasonably warm weather—the warmer the weather, the better the roofing will fit the angles of the roof.

At *B*, Fig. 191, we are showing a cant strip in the angle for the purpose of easing the bend of the roofing. Such cant strips are made by ripping 2x2's diagonally and placing them in the angles as shown. At *C* we show a cant board placed in the angle of a flat roof to ease the bend of the roofing. Such cant boards can be cut to any convenient width, and placed in the angle somewhat on the order shown.

Segmental Lattice Truss.—Fig. 192 shows a side view of a segmental lattice truss. The top chord can either be made of straight timbers as shown, or the timbers can be sawed to the proper curvature. The bottom chord is built up with straight timbers, but the joints are broken. The truss is fastened together with nails and bolts. The number and the size of them must be governed by the span

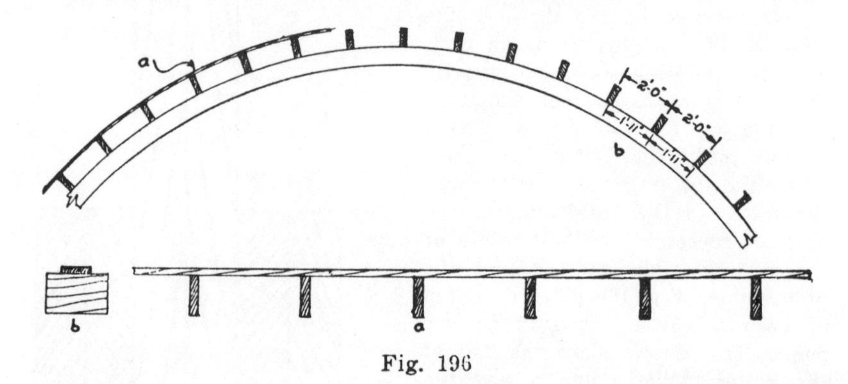

Fig. 196

and by the load the truss must support. Fig. 193 shows a cross section of the truss drawn on a much larger scale.

A diagram showing how to lay off the truss for the braces is shown by Fig. 194. Drop the dotted lines *A-D* and *B-E*, and with the compass set to one-half the span, or *A-C*, *B-C*, establish the points *D* and *E*. Divide

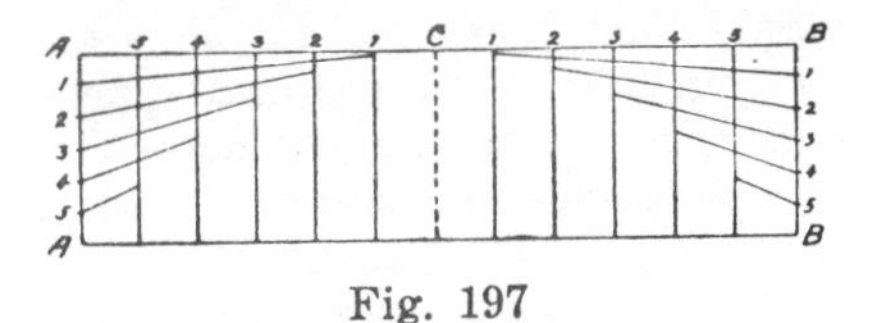

Fig. 197

the segmental chord into as many spaces as the case requires, in order to establish the points to work to. In this instance we have 0, 1, 2, 3, and 4. Now draw lines from *D* and from *E* to 0, 1, 2, 3, and 4. These lines give the location of the braces. Another method of laying off a segmental lattice truss is shown by Fig. 195. Here the lines establishing the points for the braces radiate from

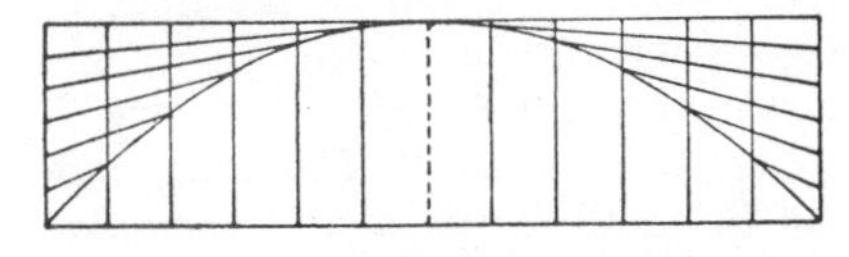

Fig. 198

a common center, as indicated by the dotted lines. The cross-lines representing the braces are then drawn in.

Spacing Joists over a Circle.—How to space the joists on a segmental truss in order to bring the top edges to a definite distance from center to center is shown by Fig. 196. The upper part shows a part of a segment, on which the joists have been spaced so that the upper edges will be 2 feet on center. At the bottom is shown, *a*, a side view of a templet for obtaining the spacing and, *b*, an end view. A 1x6, as shown, is nailed to short blocks of the joist material and sprung so it will fit the circle as at *a*, the top drawing. By measuring the spaces of these blocks on the circle,

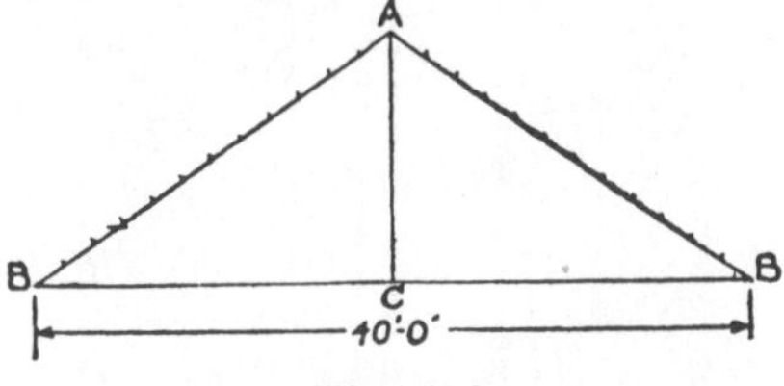

Fig. 199

we find that 1 foot, 11 inches, or 23 inches is the space necessary on the circle to make the upper edges come out to a space of 2 feet from center to center. This is clearly shown at *b*, to the upper right. See Fig. 188.

Laying Out Segmental Trusses.—In cases where it is impossible to

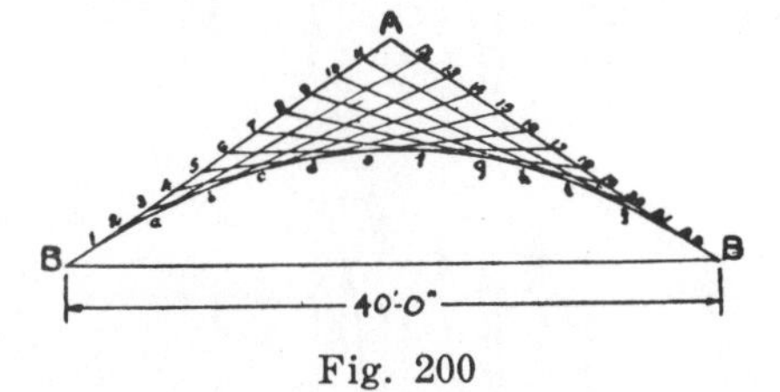

Fig. 200

use the radius for striking the segment, whether for a truss or for some other purpose, the method we are illustrating by Fig. 197 will give good results.

Draw the two horizontal lines marked *A-B*, and the perpendicular lines *A-A* and *B-B*. Now establish

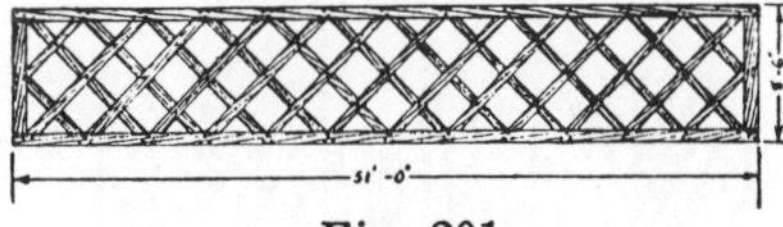

Fig. 201

the center, marked *C*, and put in the dotted line. This done, divide the spaces *A-A*, *A-C*, *C-B*, and *B-B* into the same number of parts, and number them as shown. *A-A* and *B-B* from top to bottom, and *C-A* and *C-B*

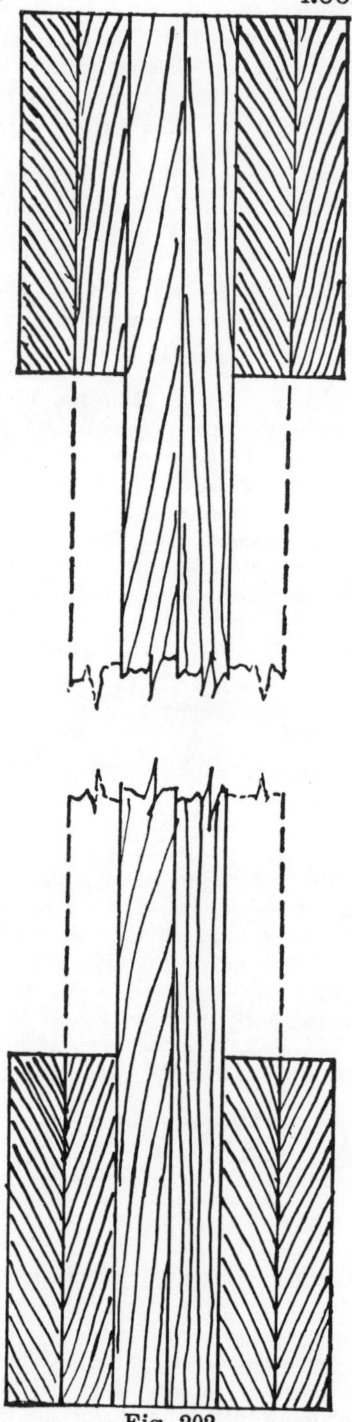

Fig. 202

from the center out. After putting in the rest of the perpendicular lines, draw a line from 1 toward *C*, both from right and left, stopping at the perpendicular lines marked 1, and from 2 toward *C*, stopping on 2, and from 3 toward *C*, stopping on 3 and so on. Having these lines, draw a curve cutting the intersections on the order shown by Fig. 198.

This method does not make a perfect circular curve, but for all practical purposes it will answer the purpose so far as segmental trusses and curved ceilings are concerned.

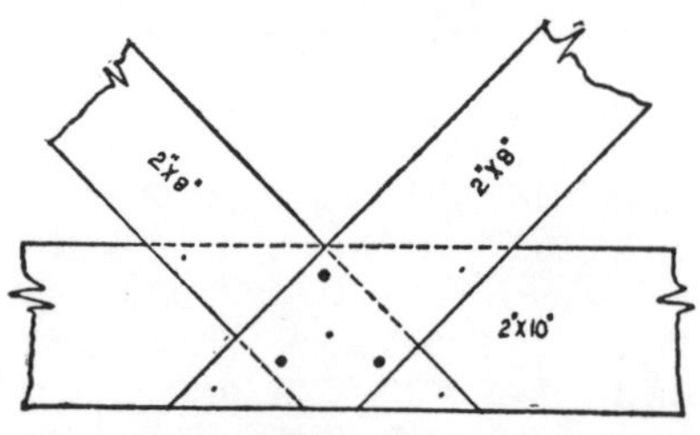

Fig. 203

Another Method for Laying Out a Segment.—Fig. 199 shows the first operations for obtaining the lines of a segment with a 40-foot chord and a rise of about 8 feet. First stretch a line from *B* to *B*, fastening each end to a nail driven at the respective points. Then obtain the center, *C*. On a line perpendicular to the chord, at *C*, measure off twice the distance of the rise, to obtain point *A*. Connect *A* with *B* and *B*, and space these distances into an equal number of equal parts—the more spaces you have, the more accurate will be your work. Drive a nail at each of these points and proceed as follows:

Stretch a line from 1 to 12, Fig. 200, another line from 2 to 13. These lines will cross at point *a*. At this point drive a nail. Remove the line from 1 to 12 and stretch it from 3 to 14, which will cross line 2-13 at point *b*, at which point drive another nail. Remove line 2-13 and fasten it to 4-15, driving a nail at *c*. Repeat these operations as often as is necessary to obtain all of the points of the arc up to point *j*. Number 1 and 22

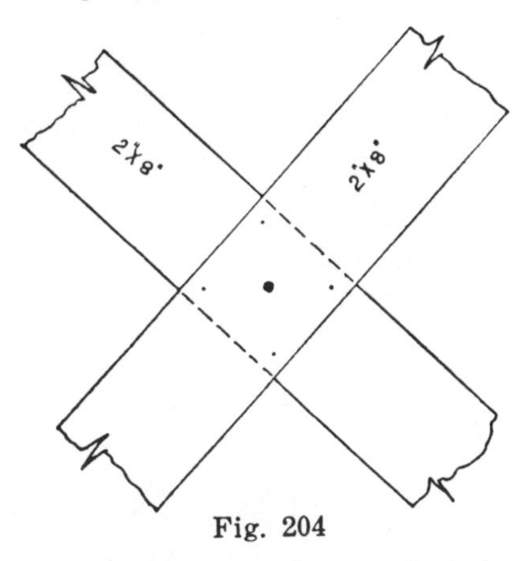

Fig. 204

are also points of the arc. By bending a straight-grained flexible board in line with these points, the arc of the segment can be marked with a pencil. When the segment is laid out on the ground, stakes must be used instead of nails.

Built-Up Lattice Truss.—Fig. 201 gives a side view (much reduced) of a built-up lattice truss. This design was used in this country long before steel. It is especially suitable for localities where steel or long timber is difficult to obtain. This truss will carry heavy loads over long spans. It was used successfully for railroad bridges, and for supporting flat roofs. The height of a lattice truss such as shown by Fig. 201, should be not less than one-eighth of the span. To determine the number of spaces for the braces, double the span and divide by the height. Or, taking the truss we have just mentioned for example, 51 times 2, divided by 8.5, equals the number of spaces for the span, or 12. The braces should be placed as nearly as possible at an angle of 45 degrees.

Fig. 202 shows a cross section of the truss, drawn to a larger scale. Both the top and bottom chords are made of four members. This makes it possible to break joints, a thing that is very important, especially in the bottom chord, which must resist a tensile strain. The braces are shown in part at the center, and the end uprights by dotted lines.

How the braces are fastened to the chords is shown by Fig. 203. The

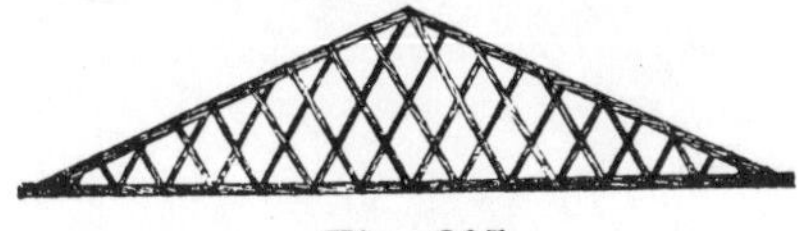
Fig. 205

small dots represent nails, and the three heavy dots show how the bolts should be placed. Fig. 204 shows how the cross joints should be nailed and bolted.

Light Lattice Trusses.—Fig. 205 shows a lattice truss that is suitable for rather short spans and light loads. This truss is subject to modifications. For instance, if it is used for supporting a gable, as of a porch, the outside can be boxed on an incline, placing the braces on the inside, as

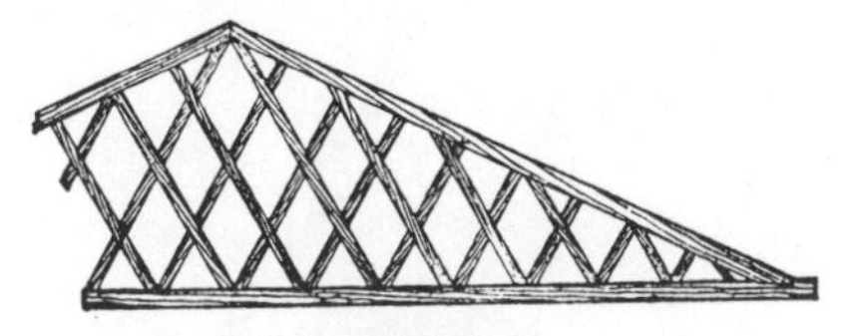
Fig. 206

shown. The spacing of the braces can be increased in numbers, or decreased, whichever the case might require.

Another Lattice Truss Construction.—Fig. 206 shows a different construction of the truss just explained. Here the rafters and the chord are nailed onto the braces on either side, whereas the braces were nailed to the rafters and the chord in the design shown by Fig. 205.

LESSON 27

DIFFERENT PITCH TRUSSES

Limited Use for Wood Trusses.—Trusses constructed of wood are not used as extensively as they used to be. Steel has an undisputed place in the building world, and will have increasingly so. But we believe that trusses constructed of wood will always be used to some extent; however, the longer spans will be taken over by steel, more nearly than the shorter ones. It is hardly probable, though, that trusses constructed of wood will completely go out of use. Their use will be limited, perhaps, to certain localities, especially localities where heavy timber is easily available, or to localities where the cost of transportation will make it expedi-

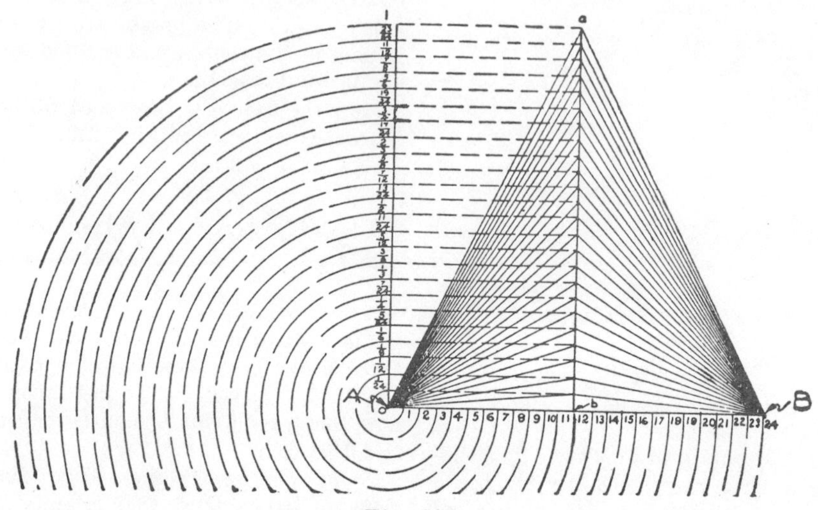

Fig. 207

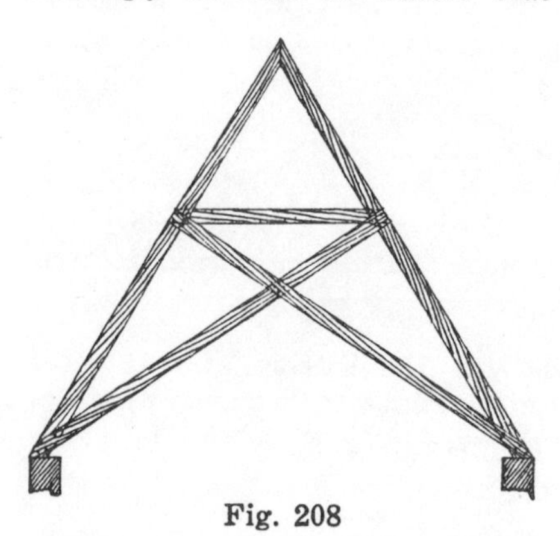
Fig. 208

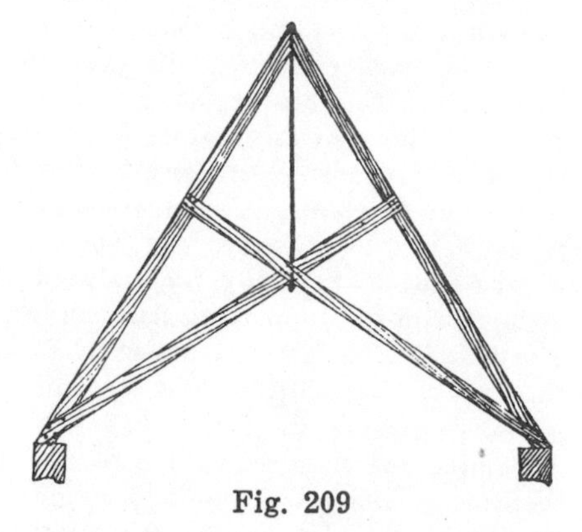
Fig. 209

ent to use wood in preference to steel. The lighter wood trusses will always be in general use. This being true, every carpenter should be informed on truss construction.

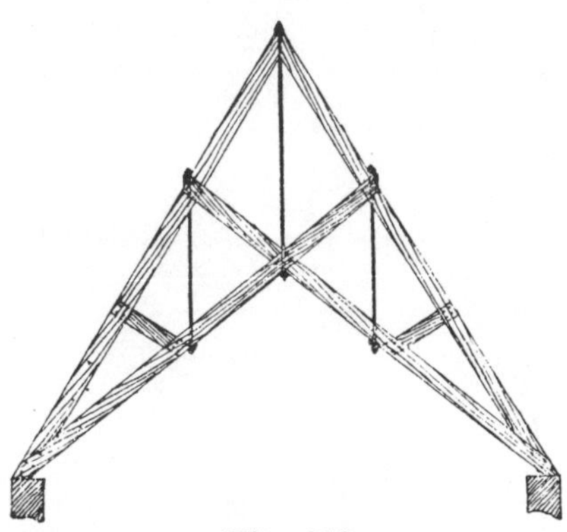

Fig. 210

Unlimited Pitches.—There are as many pitches in roof framing as there are slopes to which the rafters can be made to slope. To illustrate this we are offering Fig. 207, where a full pitch has been divided into 24 different pitches. The distance *a-b*, or a full-pitch rise, is equal to the distance *A-B*, the span. The dashed three-quarter circles (most of them

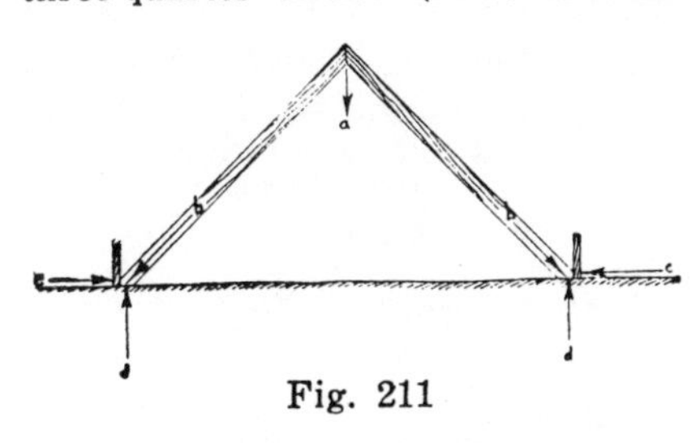

Fig. 211

have been cut off at the bottom of the illustration) have been made with a radius from one foot up to twenty-four feet. The figures, 1, 2, 3, 4, and so forth, just below the base line, represent the different radii in feet of the pitches shown. The pitches, as shown by the perpendicular line running from the center up, run from $\frac{1}{24}$ pitch up to a 1 pitch, or as it is commonly called, a full pitch. Between any two pitches shown in the diagram, an indefinite number of pitches could be placed, and therefore the pitches are unlimited.

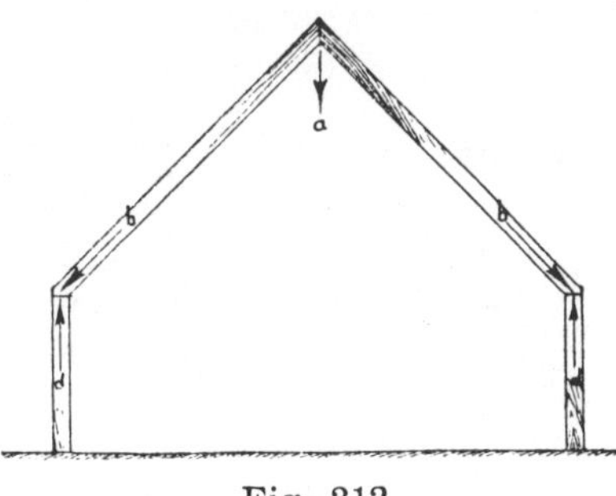

Fig. 212

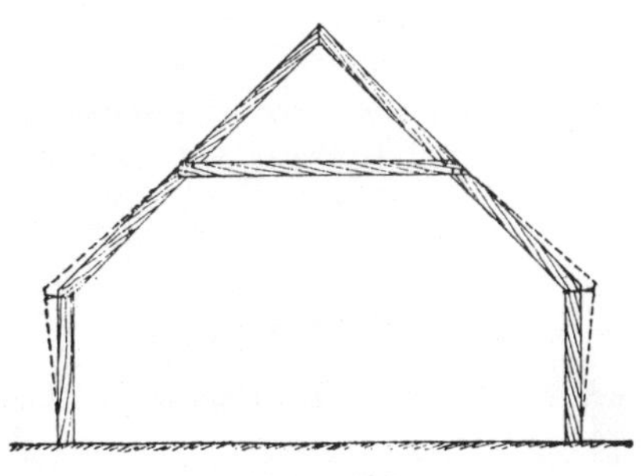

Fig. 213

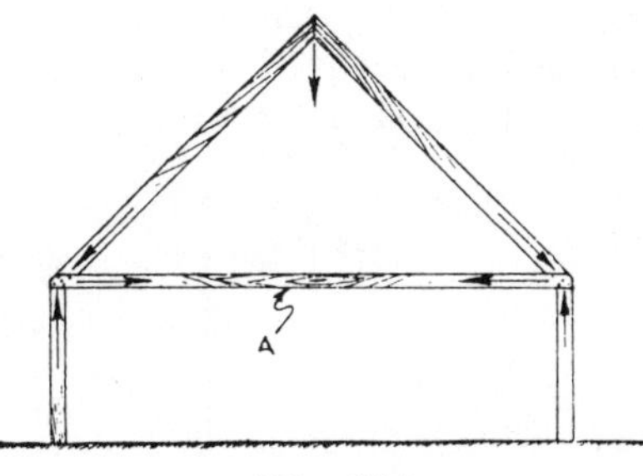

Fig. 214

Scissors Trusses.—Scissors trusses are mostly used in church roofs, in which the trusses are exposed. Figs. 208, 209, and 210 are rough designs of scissors trusses. Because of the increased tensile strain on the tie beams, they should be made of extra good material.

Counteracting Forces.—Fig. 211 shows a pair of rafters resting on the earth with a stake at either side, keeping them from spreading. The arrow at *a*, shows the direction of the force of gravity. The arrows at *b* and *b*, show how the force of gravity has been changed from its natural direction to that indicated by the arrows. The arrows at *c*, *c* show the direction of the resisting force, caused by the stakes. At *d* and *d* we show the gravity-resisting force of the earth.

Fig. 212 shows the same rafters, elevated by two posts. Here the force *c*, *c* is absent, and the weakness of the truss is apparent. The weakness in the design shown by Fig. 212 has been remedied by means of the tie beam, *A*, shown in Fig. 214. Fig. 213 shows the rafters tied together with a collar beam. The dotted lines indicate where the weak points of this construction are. Fig. 214 shows the pair of rafters properly tied together, which constitutes a substantial truss.

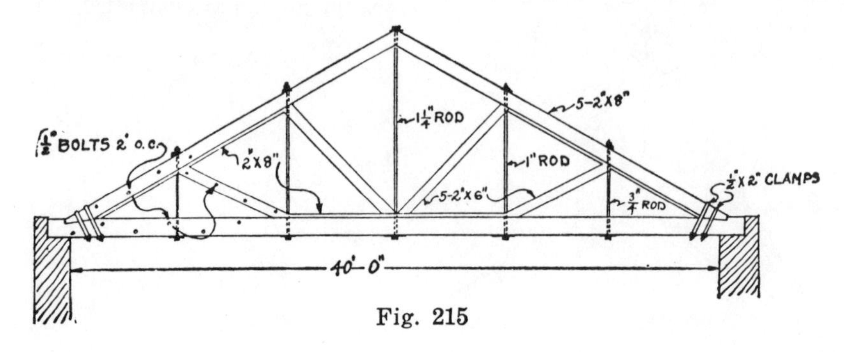

Fig. 215

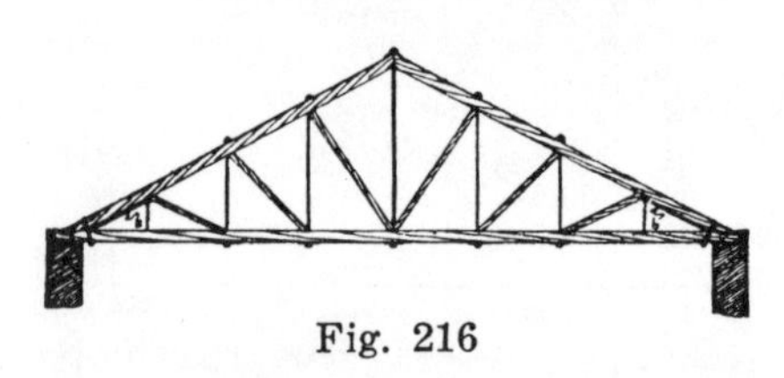

Fig. 216

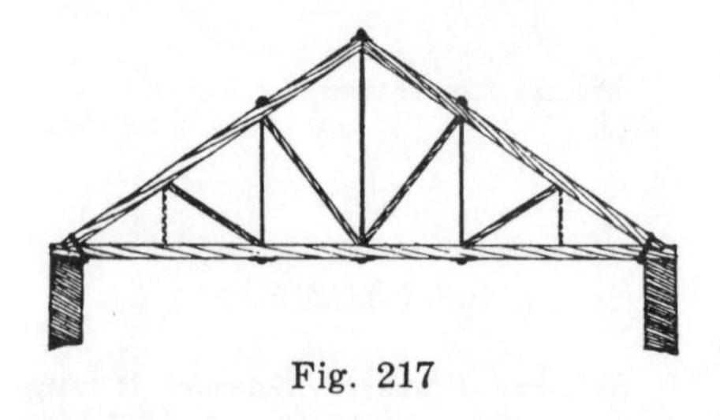

Fig. 217

Fig. 218

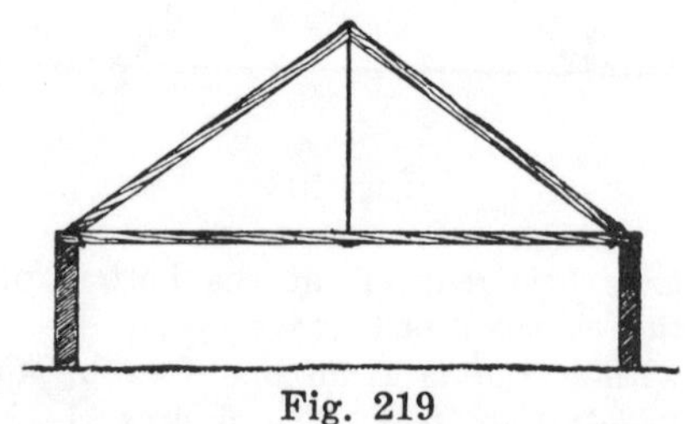

Fig. 219

Built-Up Truss.—Fig. 215 shows a drawing of a five-member built-up truss. The clamps holding the foot of the rafters to the chord should be slightly sunk into the chord to prevent slipping. The beveled washers at the upper ends of the rods should also be sunk into the rafters for the same reason. The important thing about building a truss is to give all butt joints a full bearing.

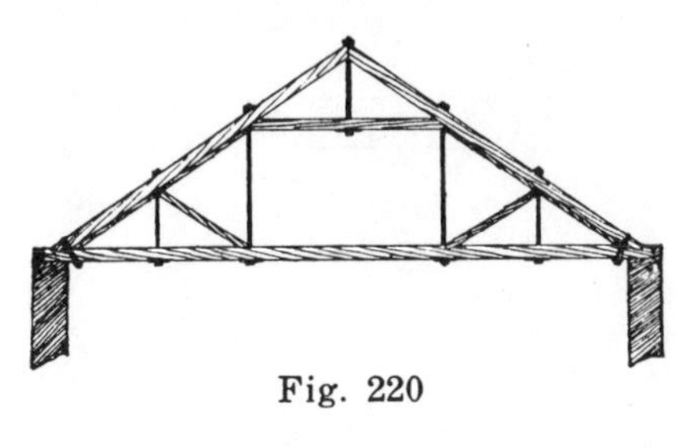

Fig. 220

Queen Truss.—Fig. 216 shows an eight-panel queen truss. Attention should be called to the two pieces of timber placed under the rafters in the end panels, pointed out at *a* and

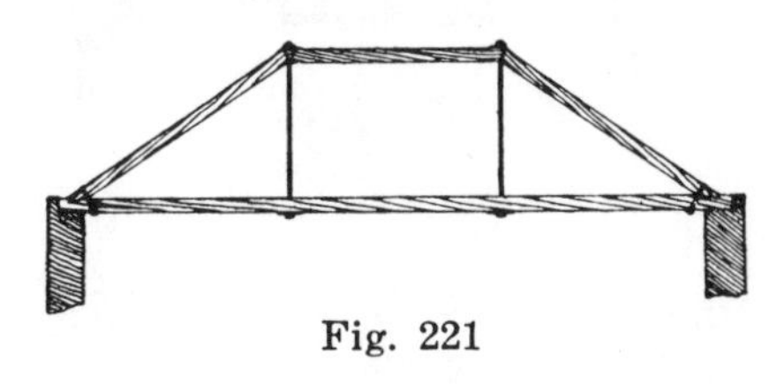

Fig. 221

at *b*. These pieces are to prevent the two short struts from slipping on the rafter. A six-panel queen truss is shown by Fig. 217. When the rafters

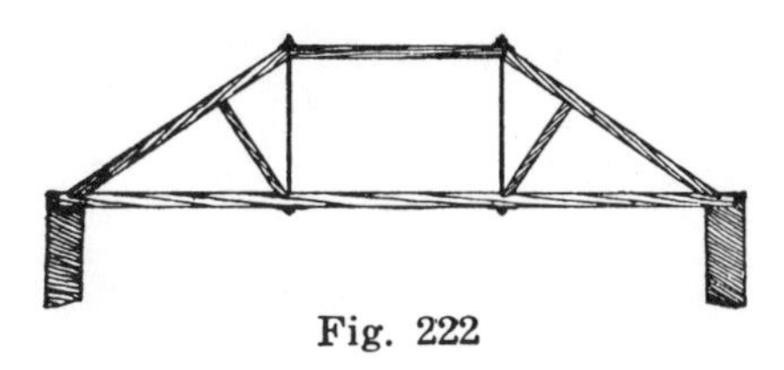

Fig. 222

of a truss are longer than 24 feet, there should be at least two struts supporting each rafter, as shown by the illustrations. If the tie beam must carry a ceiling or some other load, two additional rods should be placed as indicated by the dotted lines.

King-Rod Truss.—Fig. 218 gives a king-rod truss with two braces,

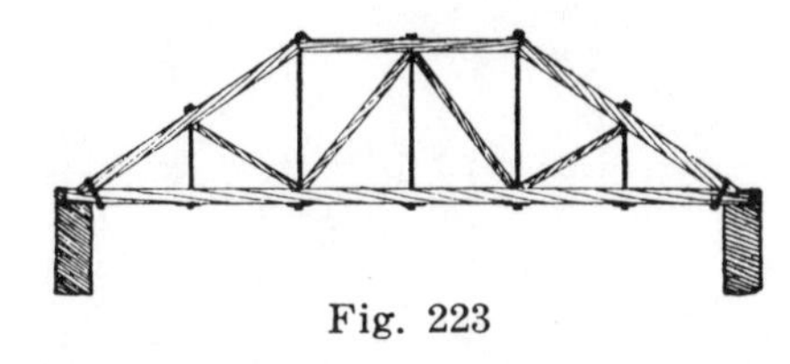

Fig. 223

marked *B* and *B*. The dotted lines show where additional rods should be placed in case the tie beam is not strong enough to carry the ceiling load. A simple king-rod truss is shown by Fig. 219. (When a post is used it is called a king-post truss.)

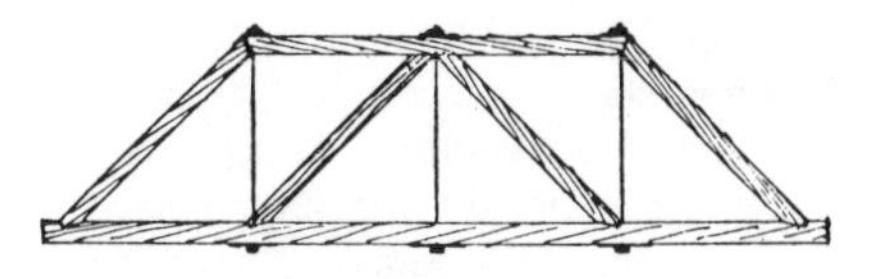

Fig. 224

Combination of Queen- and King-Rod Truss.—A form of truss that is neither a king-rod nor a queen-rod truss, but more nearly a combination of the two, is shown by Fig. 220. This truss is suitable when it is necessary to have a clear space through the center, as shown. This unobstructed space might be needed for doors, or in some cases for attic room.

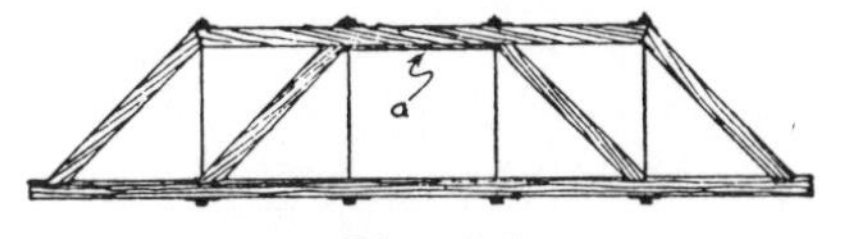

Fig. 225

Queen-Rod Truss.—Fig. 221 shows a simple queen-rod truss. This truss is not as firm as the king-rod truss, because of the horizontal piece joining the rafters at the top. However, there are many places where this truss is more suitable than the king-rod truss. Adding two struts, or

Fig. 226

braces, as shown by Fig. 222, will stiffen the truss a great deal. In extra long spans a construction such as is shown by Fig. 223 is advisable.

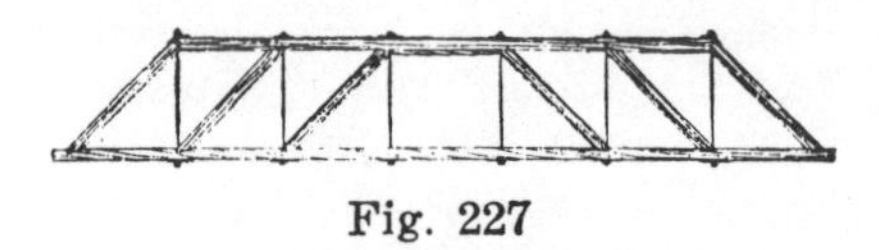

Fig. 227

Howe Trusses.—A four-panel Howe truss is shown by Fig. 224, and Fig. 225 shows a five-panel truss. An extra piece, marked *a,* is added here to prevent the two inside struts from slipping on the upper horizontal piece. Fig. 226 shows a six-panel Howe truss, and Fig. 227 shows the same kind of truss, with seven panels. These trusses are suitable for supporting bridges, roofs and floors.

Fig. 228 shows the end construction of a Howe truss for a rather flat roof. The key at *a*, and the clamp at *b*, should be noticed as important fea-

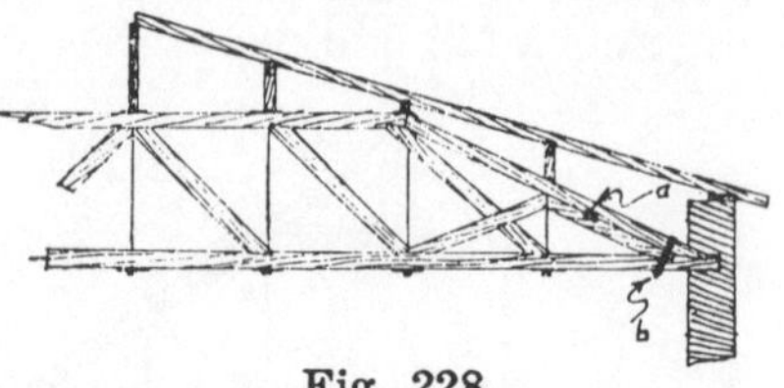

Fig. 228

tures. Another form of end construction for a long Howe roof truss is shown by Fig. 229. The keys at *a*, the clamp at *b*, and the reinforcing

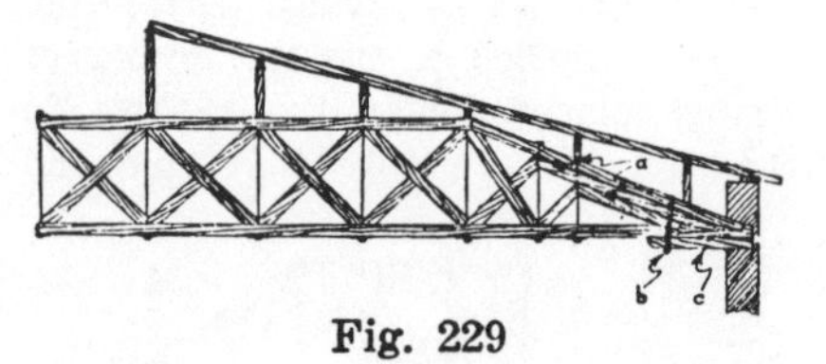

Fig. 229

block at *c* are pointed out as matters of import. The panels of this truss have two struts, as shown, forming an X when in place.

LESSON 28

HEAVY TIMBER ROOF DESIGNS

The Broad Ax.—America would not be what she is today, if it had not been for the part played in her pioneer days by the broad ax. This tool

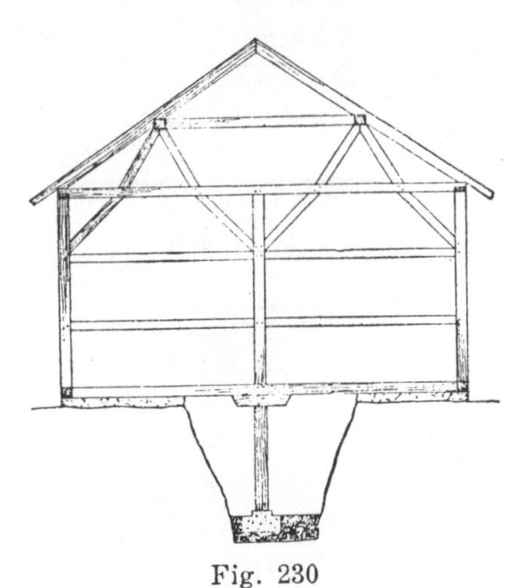

Fig. 230

(obsolete today, excepting perhaps in isolated localities) made it possible for settlements to spring up along

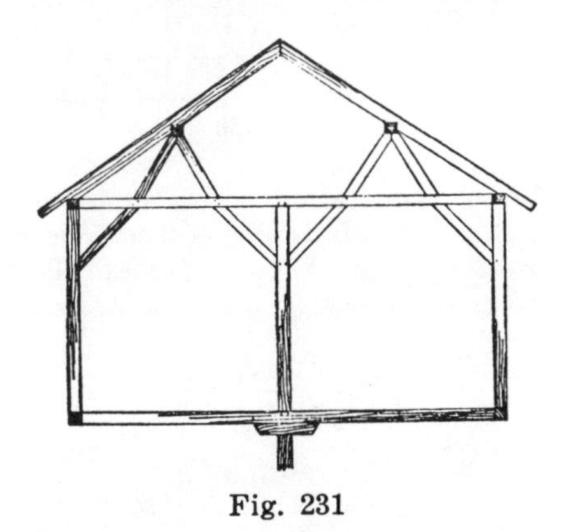

Fig. 231

the frontiers, because with it the pioneers were able to build substantial buildings for shelter, and for protection against Indian raids. For a well-built log house in those days was a veritable fortress, often saving the lives of whole settlements.

Heavy Timber Framing.—Heavy timber framing, as a rule, is not altogether a thing of the past. In rural

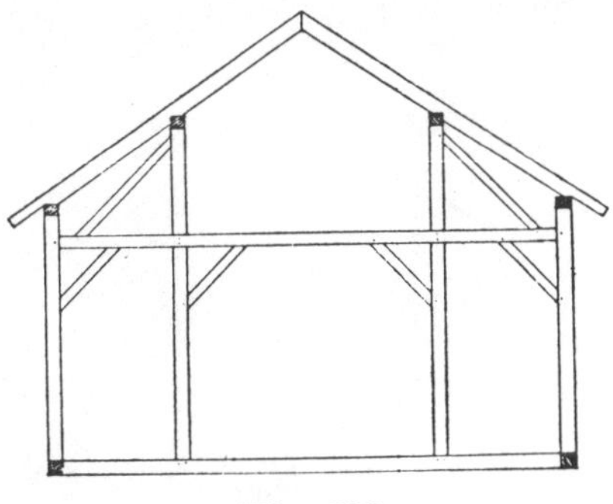

Fig. 232

districts where timber is easy to obtain, heavy timber still holds an important part in the construction of barns and other farm buildings. A short time after the first World war, I worked on two especially large hay barns, which would have to be classed

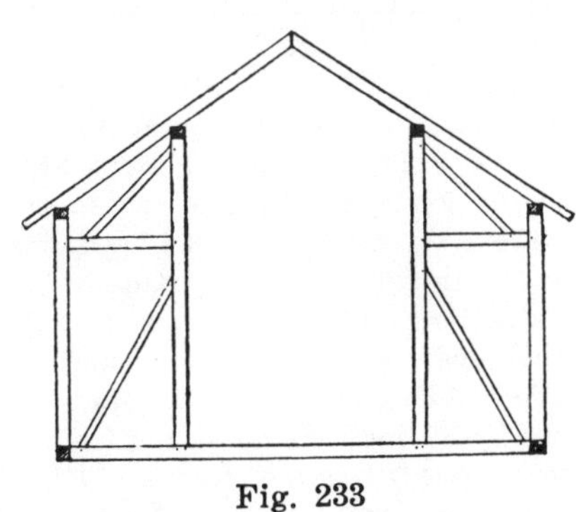

Fig. 233

as heavy timber barns. The first carpenter work that I did was on heavy timber barns that my father built. The first experience that I had work-

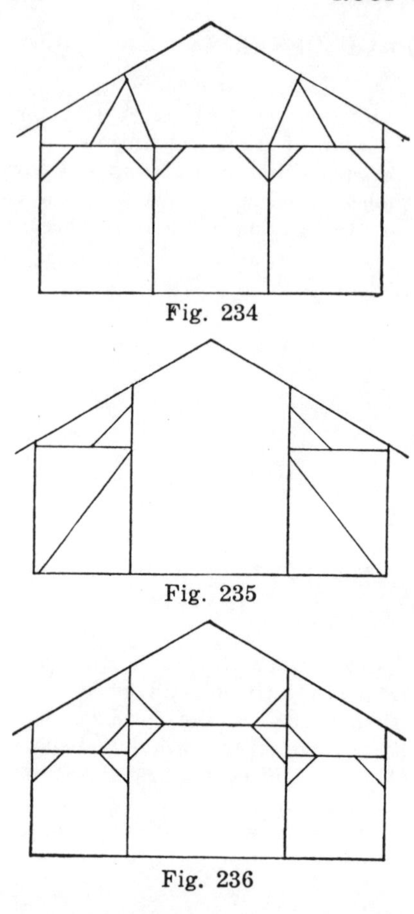

Fig. 234

Fig. 235

Fig. 236

ing for contractors was in heavy timber framing. I was the apprentice on the job and it fell to me to run the boring machine. The head carpenter did the mortising. One of the contractors laid off the work, and the other one framed the braces and the rafters. It was for a barn about 40 feet by 80 feet. Some of the timbers were hewed, but the greater part of them were sawed at a local saw mill. This barn was probably the last heavy timber barn that was built in that particular community.

Sizes of Timbers.—The different sizes of timbers that were used, as I remember them, were somewhat on this order: Sills, 10x12 or 12x12; posts and girders, 10x10 or 10x12; plates, 6x8 or 8x10; purlins, 8x10 or 10x10; purlin posts, 6x8 or 8x8;

Fig. 237

Fig. 238

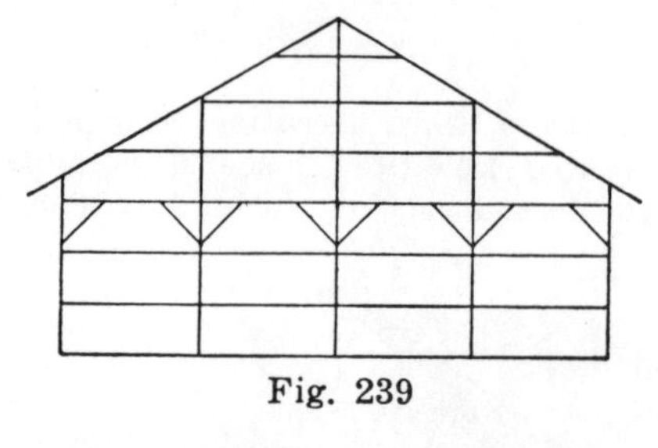

Fig. 239

braces, 4x4; rafters and collar beams, 2x6 or 2x8. Mortises and tenons were mostly 2 inches by 4 inches. The holes for the pins were bored about

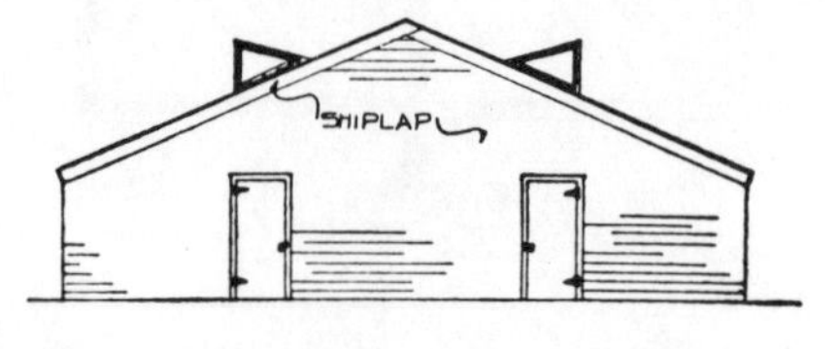

Fig. 240

2 inches from the shoulder. The hole in the tenon was bored about 1/8 of an inch or 3/16 of an inch closer to the shoulder than the hole in the main

timber. The purpose of this was to make the pin draw the joint up tight when it was driven. This is called drawboring.

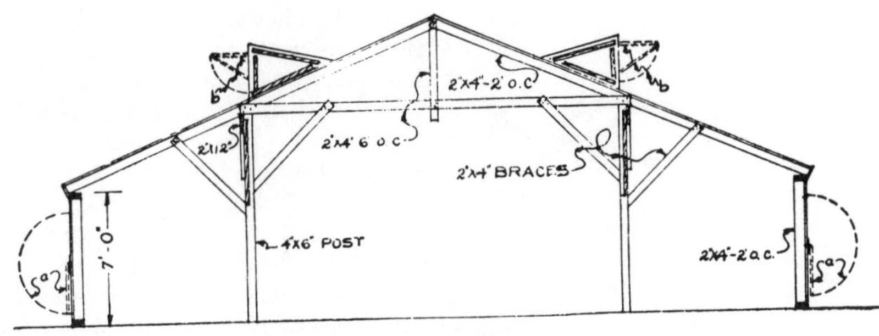

Fig. 241

Boxing.—Boxing is a term that means about the same as sizing. The purpose is to make the timbers, where the joints come, exactly the same in size, otherwise it would cause trouble.

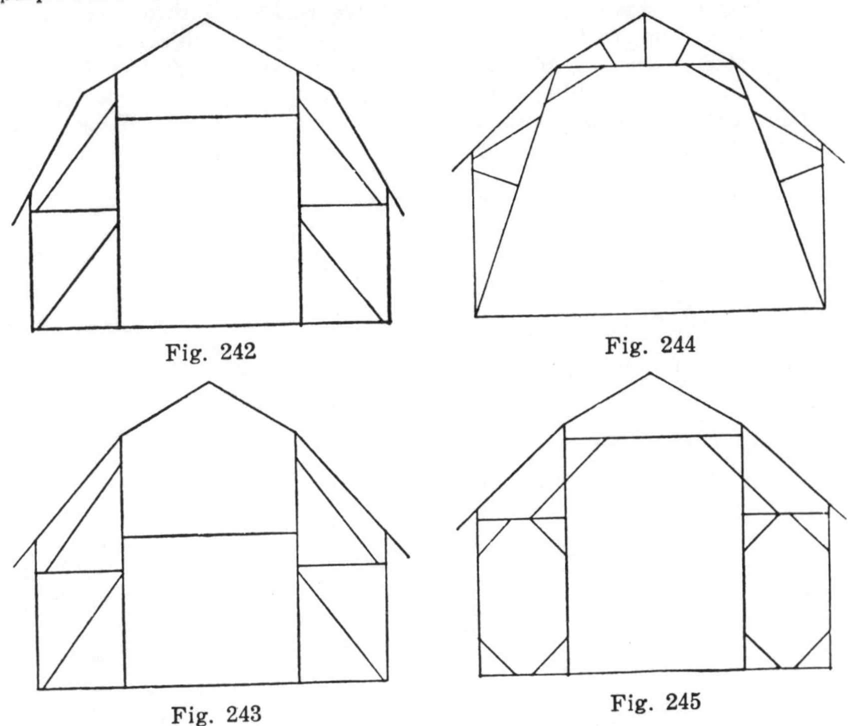

Fig. 242

Fig. 244

Fig. 243

Fig. 245

Heavy Timber Roof Constructions.—Fig. 230 shows the end construction of a heavy timber barn. The roof construction is simple but substantial. Fig. 231 shows the construction of one of the inside sections, or bents, as they were called. Fig. 232 gives a construction of a bent, that is suitable for a barn in which a track for a hay fork will have to be installed. Fig. 233 shows a construction that leaves the center part entirely clear.

Diagrams of Roof Designs.—Figs. 234, 235, and 236 are diagrams showing simple designs for heavy timber

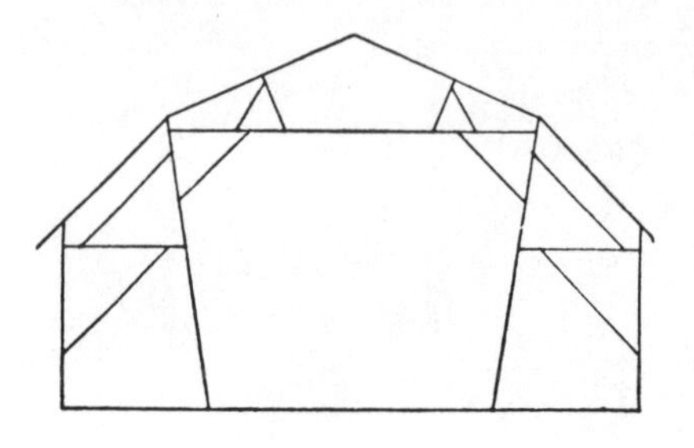

Fig. 246

barns, especially the roof constructions. Fig. 237 shows a design in which the greater part of the roof is supported by means of a truss. This construction should be used with caution. Fig. 238 shows a design for one end of a barn with a double pitch roof, and Fig. 239 shows a little different design.

Temporary Roof Constructions.—Fig. 240 gives an end view of a temporary building with a low pitch roof. Fig. 241 shows an inside section of the same roof, on a larger scale, giving the sizes of the timbers in figures. The dotted lines at *b* and *b* show how the hinged sash are opened to provide ventilation for the interior. The dotted half circles at *a* and *a* show how hinged shutters are opened to provide more ventilation.

Gambrel Barn Roof Designs.—Fig. 242 is a simple design, that is at the same time substantial, and Fig. 243 gives a slightly different design. Fig. 244, 245, and 246 are a little more complicated designs.

It should be remembered that all of the roof designs that are shown here are given subject to modifications. Every roof framer must adjust the particular roof design that he chooses to fit into the situation that he has before him.

LESSON 29

CORNICES

Styles of Cornices.—The styles of cornices, especially house cornices, are as various and go from extreme to extreme as much as the styles of women's hats. I can remember when

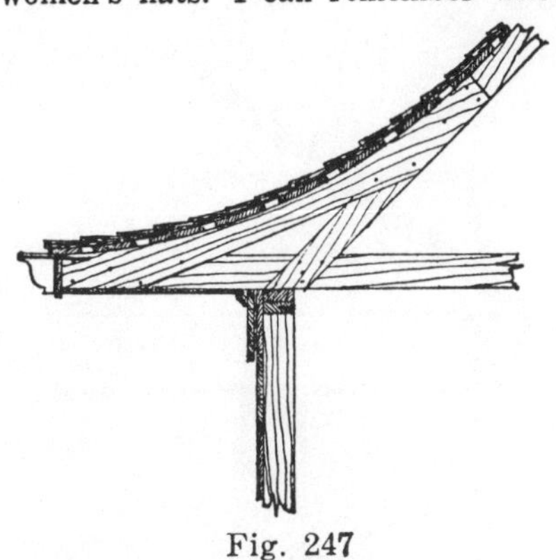

Fig. 247

women wore hats with such wide brims that under many circumstances they had to tilt the head in order to pass threatening objects without rubbing the rim of the hat against them.

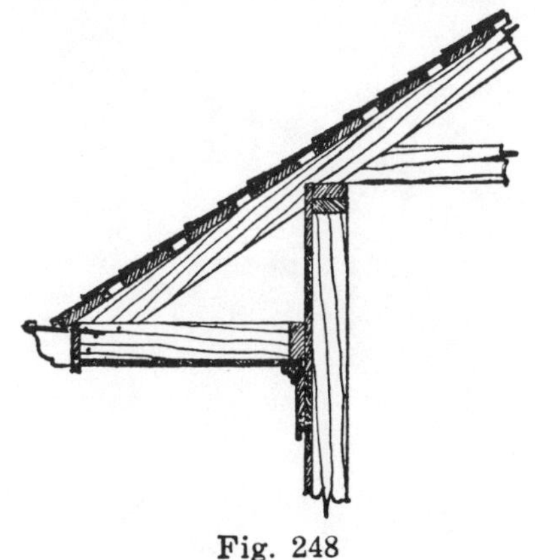

Fig. 248

Even passing through doorways had to be accomplished with a great deal of care. If the lady passed too close to one side or the other, there was danger of her hatbrim bumping against the jamb. The only safe way, if she wanted to hold her head erect,

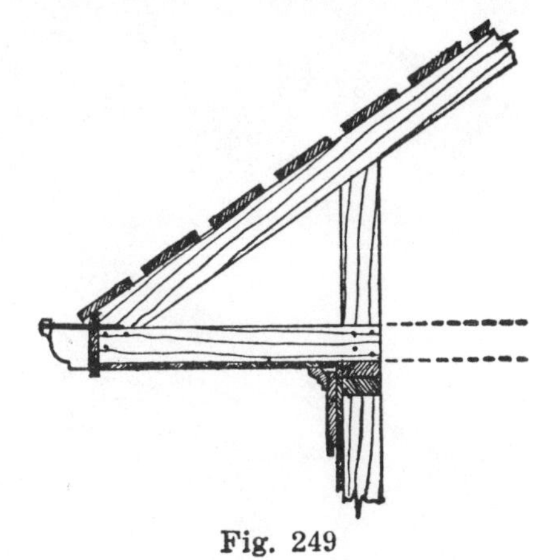

Fig. 249

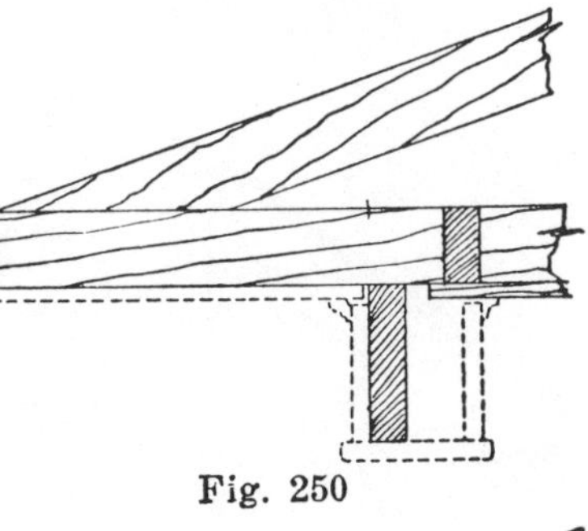

Fig. 250

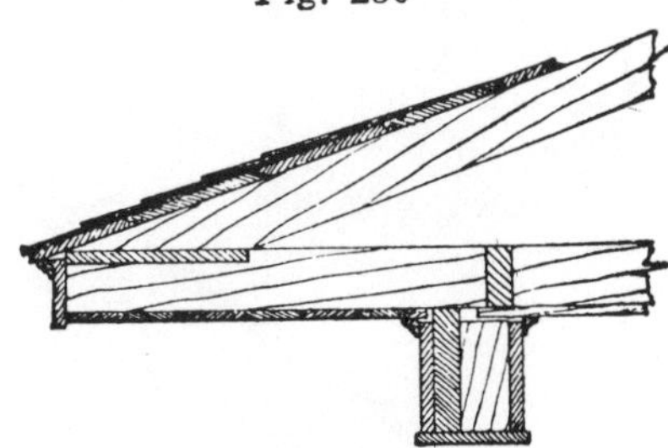

Fig. 251

was to find the center of the doorway, and carefully toe the mark. From this extreme the styles of women's hats went to the other extreme, which was a sort of brimless hat, if it could be called a hat at all. The tighter they could make the hat fit the head, the more nearly they seemed to be in style. With this close-fit style of hat, a bump on the hat meant a bump

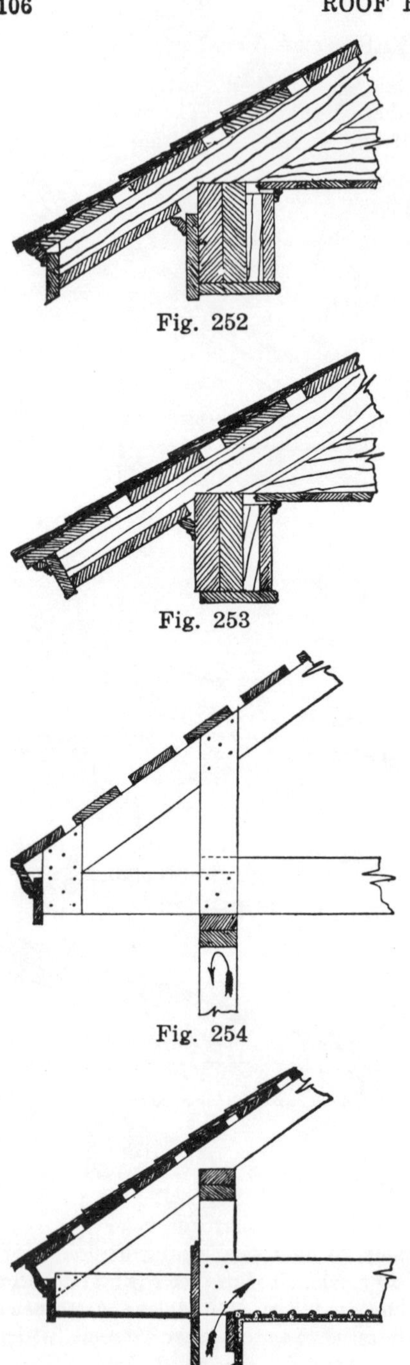

Fig. 252

Fig. 253

Fig. 254

Fig. 255

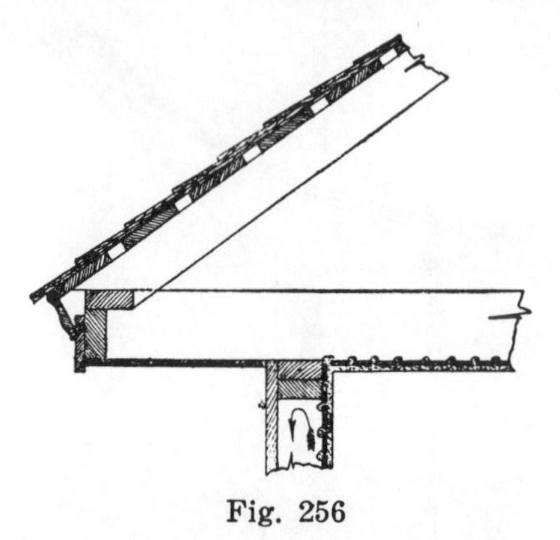

Fig. 256

on the head. In much the same way it was with cornices for houses. There was a time when especially wide cornices were so much in demand that they seemed to be not only in style,

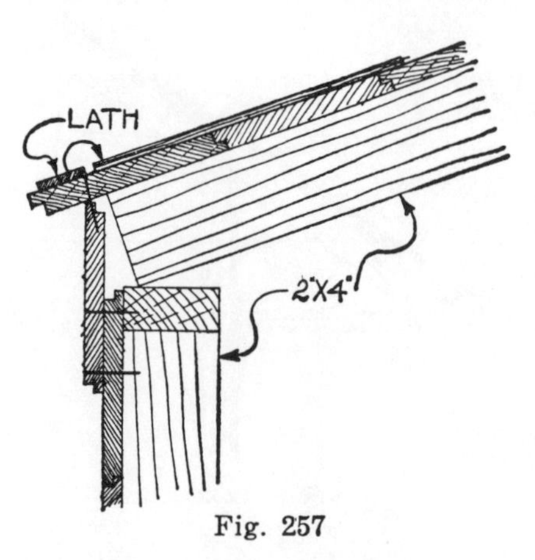

Fig. 257

but a real necessity. In those days the dehorned cornices were rarely used. But just as the women with their hats, we have gone from the extremely wide cornice that made it seem as if the porch extended all the way around the house, to the extremely narrow cornice that makes the house look, not only bare-headed, but bald-headed as well. I will discuss the different kinds of cornices from a practical standpoint as we take up the illustrations.

In this and the next lesson I am giving a big variety of cornice designs, so that the roof framer will have many kinds of cornices to choose from.

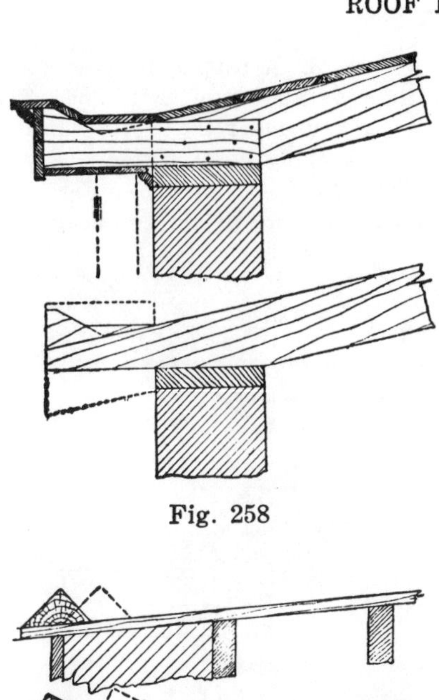

Fig. 258

Fig. 259

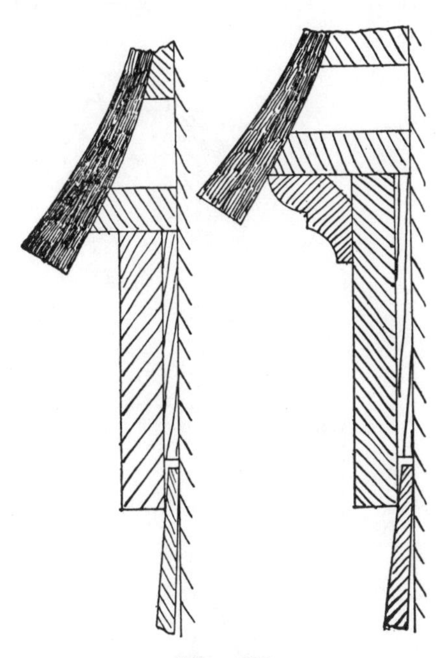

Fig. 261

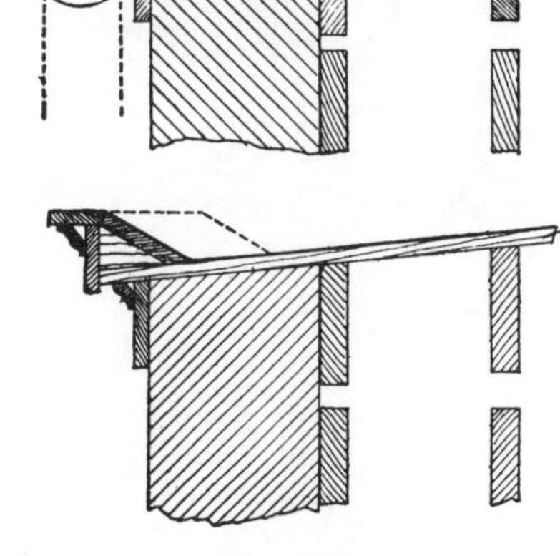

Fig. 260

Wide Cornices.—Fig. 247 shows a swayed cornice, that is, the upper part of the cornice is swayed. This style of cornice was much in demand in the early part of the twentieth century. I have seen such cornices much wider than what is shown by the illustration. Fig. 248 shows a sort of figure-4 cornice construction. Fig. 249 shows a different way of constructing such a cornice. This design raises the cornice and the roof above the plate line. The crown-mold gutters should be noted, and the two designs compared.

Closed Porch Cornices.—Fig. 250 shows the construction of a closed porch cornice, only partly completed. The ceiling plancher is indicated by dotted line. The finishing of the chord is also shown by dotted lines. Fig. 251 shows a similar cornice that is completed. In this design the rafter rests on a toe board. The two designs should be compared and

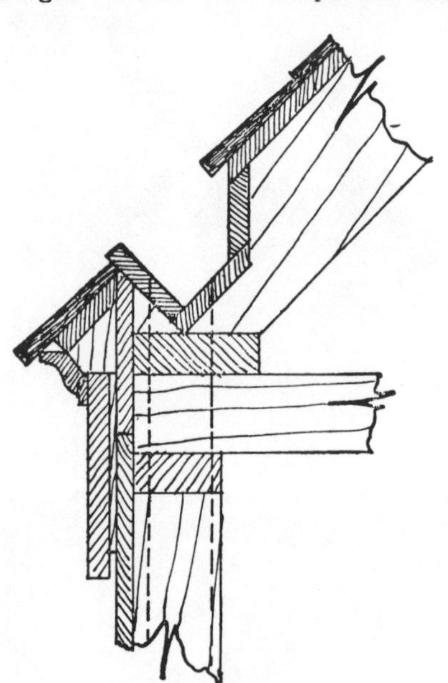

Fig. 262

studied. Fig. 252 shows a closed porch cornice together with a design for the porch chord. A little different porch cornice construction is shown by Fig. 253. The design for the chord is also different.

House Cornices.—Fig. 254 shows the rough work of a house cornice, which is on the figure-4 order. The bent arrow under the plate indicates that this is a fire-checking construction. Fig. 255 shows a similar house cornice. Here the arrow indicates a

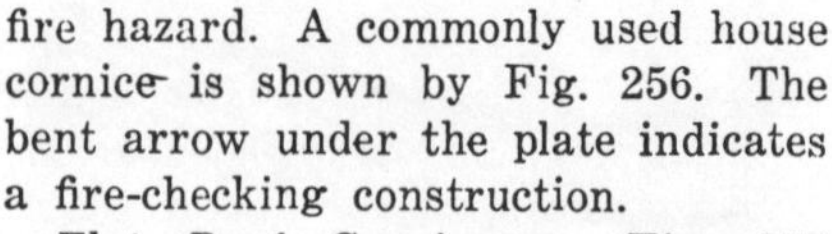

fire hazard. A commonly used house cornice is shown by Fig. 256. The bent arrow under the plate indicates a fire-checking construction.

Flat Roof Cornices. — Fig. 257 shows a cornice of a very cheap con-

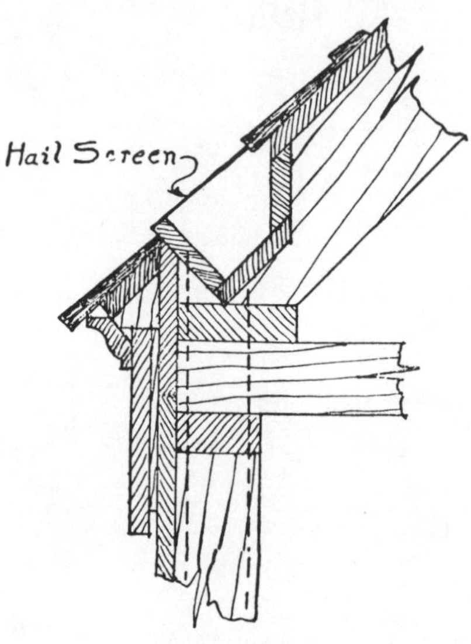

Fig. 263

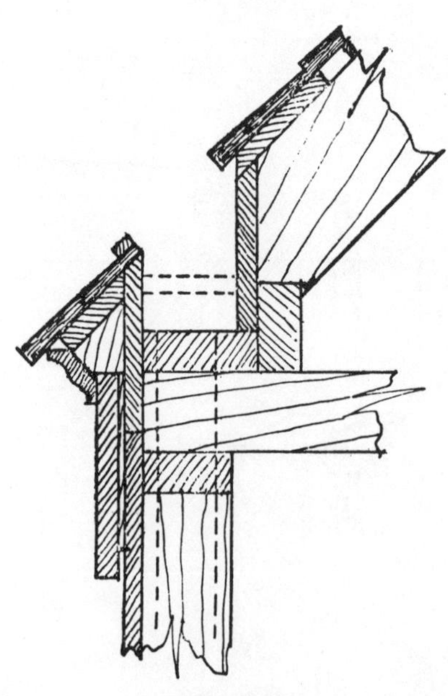

Fig. 264

struction for temporary buildings. Fig. 258 shows two designs of a cornice and gutter for a flat roof. Fig. 259 shows three different planted-on gutters for a flat roof, which also answer for the cornice, if it can be called that. Fig. 260 shows two gutter designs for flat roofs that also answer for the cornice.

Dehorned Cornices. — Fig. 261 shows two cornice designs suitable for a turret roof, or for any kind of rather steep roof. The sway shown can be modified so as to show straight shingles. Fig. 262 shows a cornice in connection with a gutter that has a concealed downspout. The dotted lines indicate the downspout. A similar cornice with a concealed downspout for the gutter is shown by Fig. 263. The hail screen cover for the gutter is a feature that should be noted. This prevents leaves from lodging in the gutter. Another cornice design with a different gutter, that also has a concealed downspout, is shown by Fig. 264. A simple design for a dehorned cornice with a simple gutter is shown by Fig. 265. Fig. 266 shows two designs of cornices with gutters made of wood. Wooden gutters, if kept well painted, will give good service and a pleasing appearance. A cornice with a sort of box gutter is shown by Fig. 267. The dotted lines indicate the high point of the gutter bottom.

Fig. 265

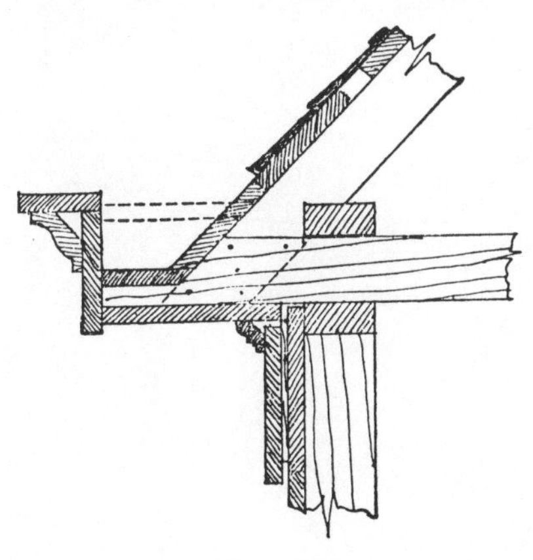

Fig. 267

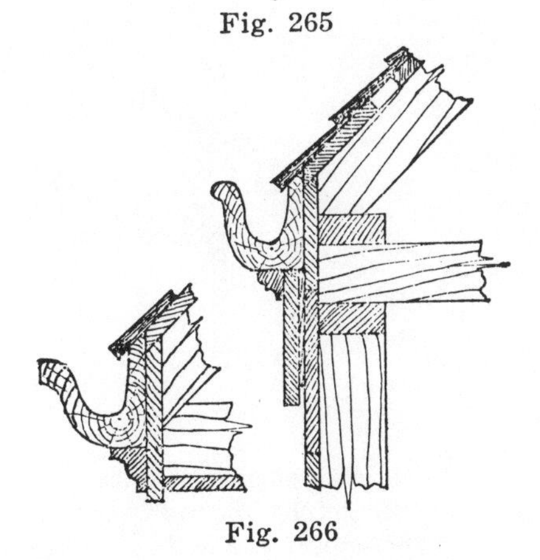

Fig. 266

LESSON 30

MORE CORNICES

The cornice can either make or mar the appearance of a building, and for that reason the designer should study the whole matter from different angles before he makes a final decision. A wide cornice costs considerably more in labor and material than a narrow one. It gives the outside of the building a certain protection against the weather, which is not a perfect protection. The upper part of the outside is protected against the elements, but it catches dust, which the rain can not wash off. The lower part is unprotected by the cornice, and when it catches dust it is washed off by the rain. Consequently the upper part of the outside has undamaged paint, but is dirty,

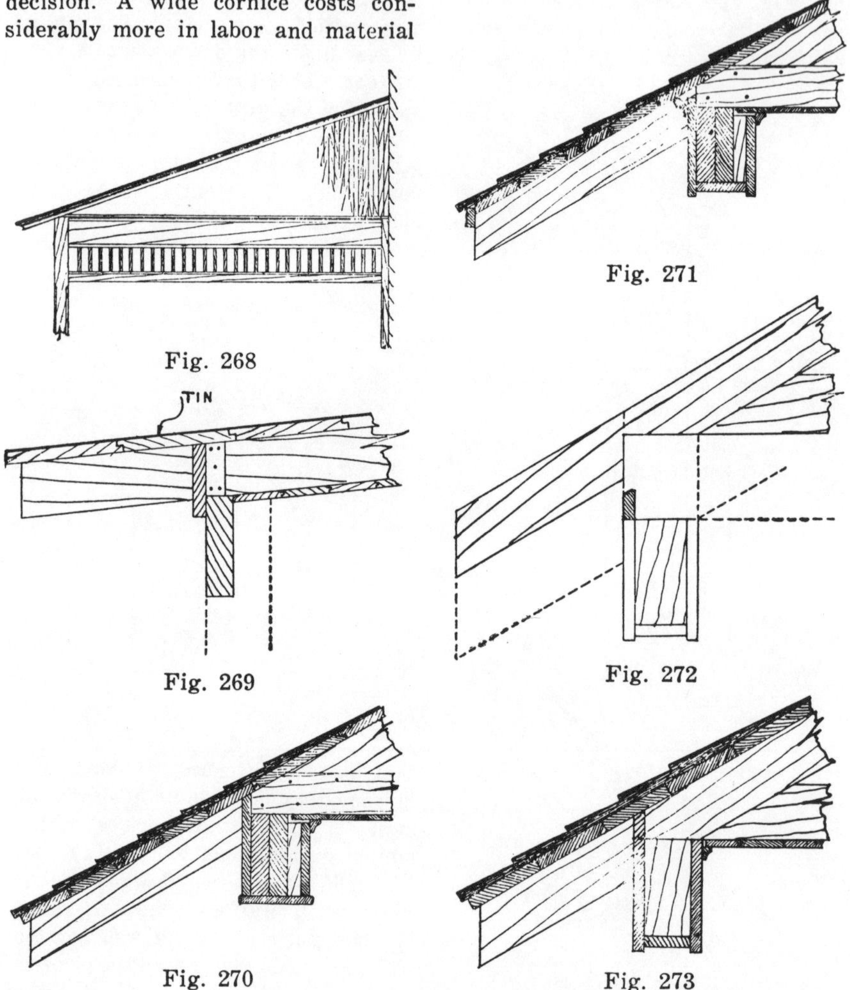

Fig. 268

Fig. 269

Fig. 270

Fig. 271

Fig. 272

Fig. 273

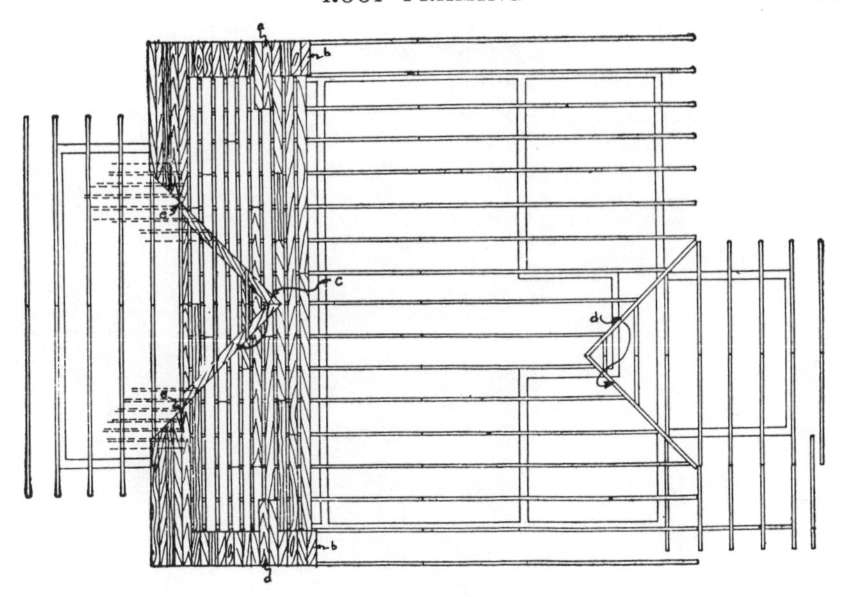

Fig. 274

while the lower part is clean with paint showing deterioration.

The dehorned cornice costs little, comparatively speaking, in labor and

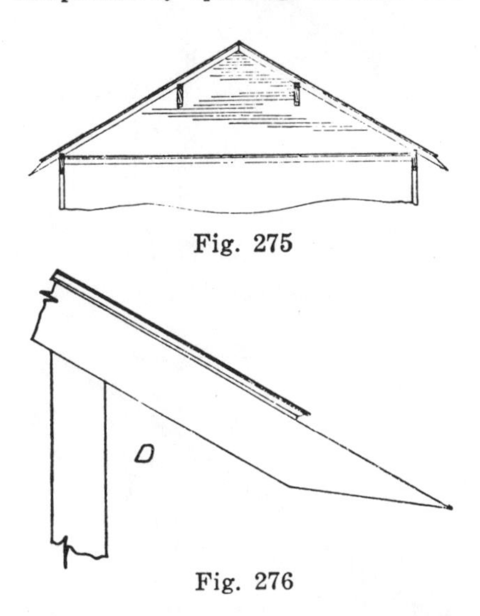

Fig. 275

Fig. 276

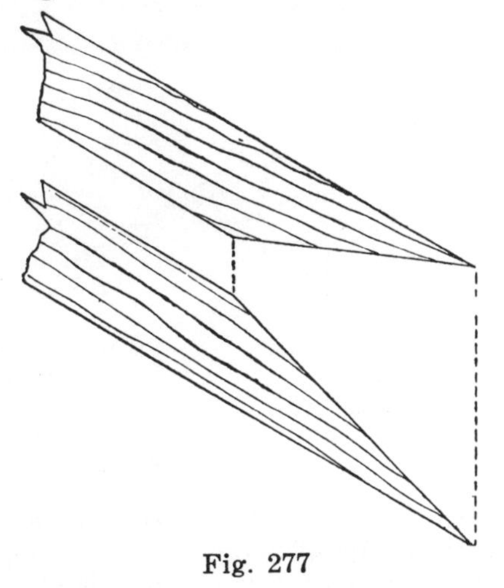

Fig. 277

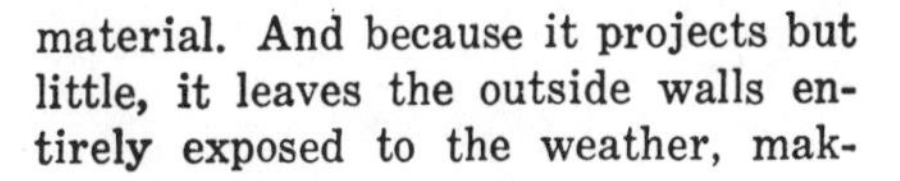

material. And because it projects but little, it leaves the outside walls entirely exposed to the weather, making them uniformly the same in appearance.

Open Porch Cornices. — Fig. 268 shows an end view of a simple lean-to porch roof with a simple cornice. Fig. 269 shows another simple porch cornice. Fig. 270 shows a design of a porch cornice together with the construction and finish of the chord.

A somewhat different design is shown by Fig. 271. Fig. 272 shows a porch rafter about to be placed in position. The dotted lines show the bottom lines of the rafter and ceiling joist,

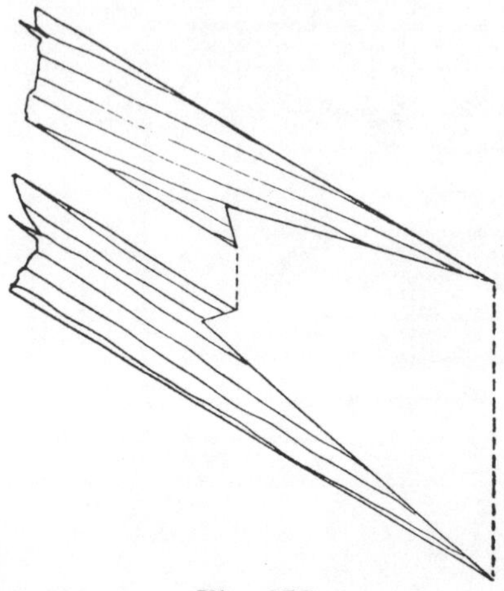

Fig. 278

when in position. This construction put together and finished is shown by Fig. 273.

Roof Plan.—Fig. 274 gives a roof plan. To the left is shown a method of joining a secondary roof to the

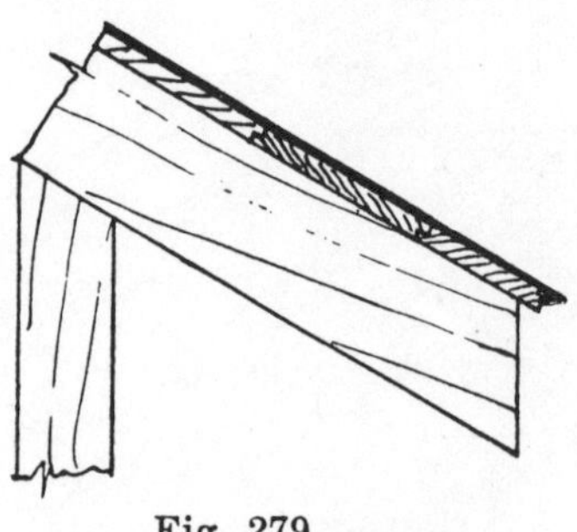

Fig. 279

main roof without the use of valley rafters. In place of valley rafters, we have valley boards. These boards, which are pointed out at *c* are nailed on the sheeting of the main roof; and to them the jack rafters of the secondary roof are nailed. When this method is used, the common rafters of the main roof all run down to the plate. Those under the secondary roof need not extend farther than to the plate.

The dotted lines show how the secondary, or porch roof should be

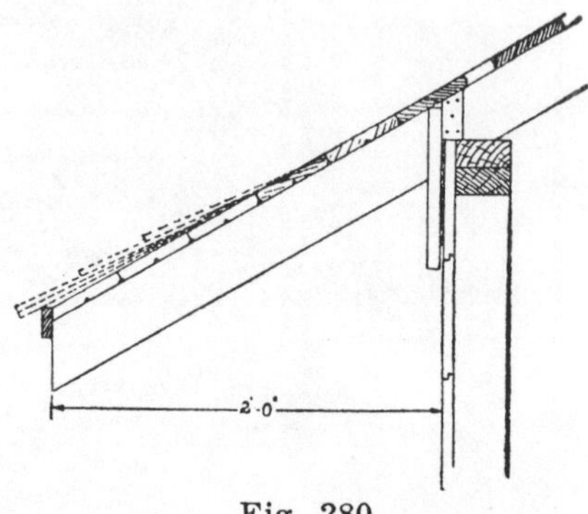

Fig. 280

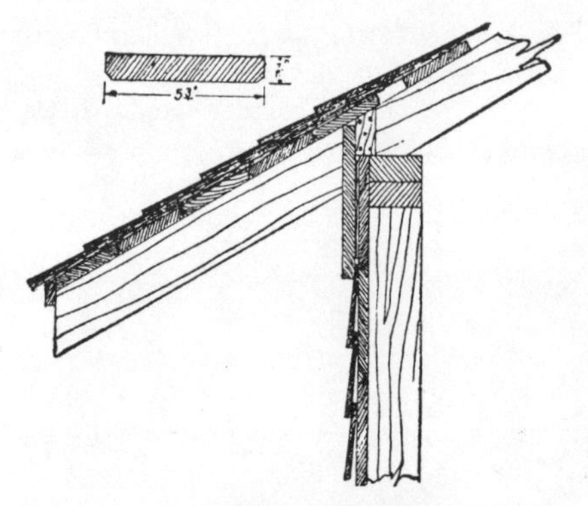

Fig. 281

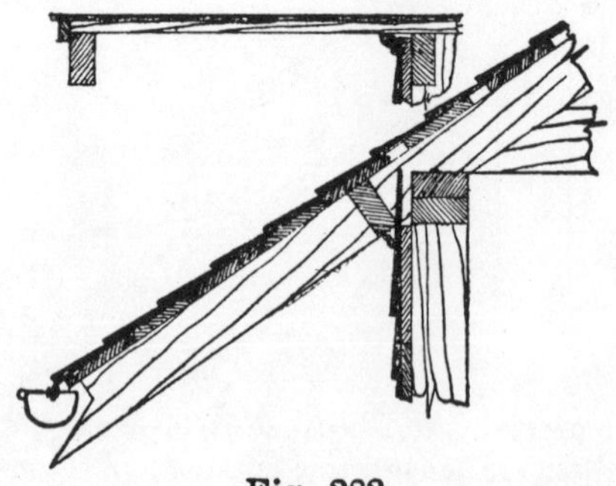

Fig. 282

sheeted in order to receive the cornice of the main roof. The points where the porch roof begins to extend

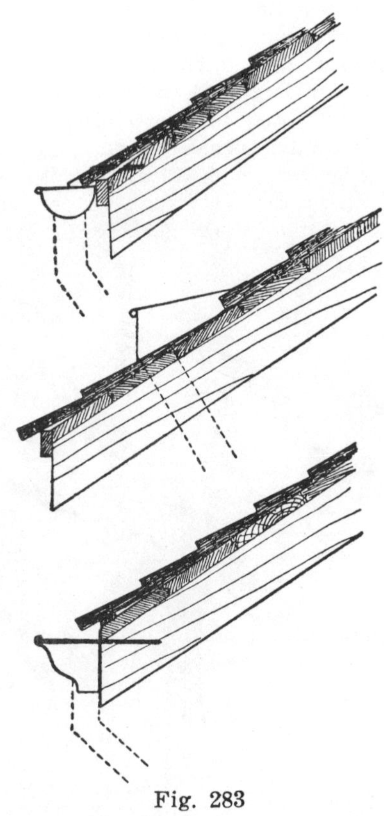

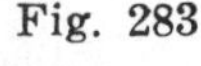

Fig. 283

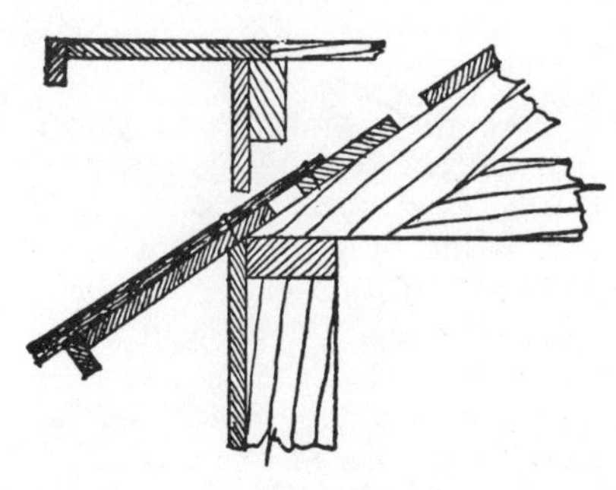

Fig. 284

over the main roof are shown at *e, e*. At *a, a* are shown how at intervals the ceiling boards for the gable cornices extend in to the second rafter, thus supporting the verge rafters. At ***b, b*** are shown the short ceiling boards for the gable cornices. The secondary, or kitchenette roof, shown to the right, should be joined to the main roof just like the porch roof is joined. However, we are showing valley

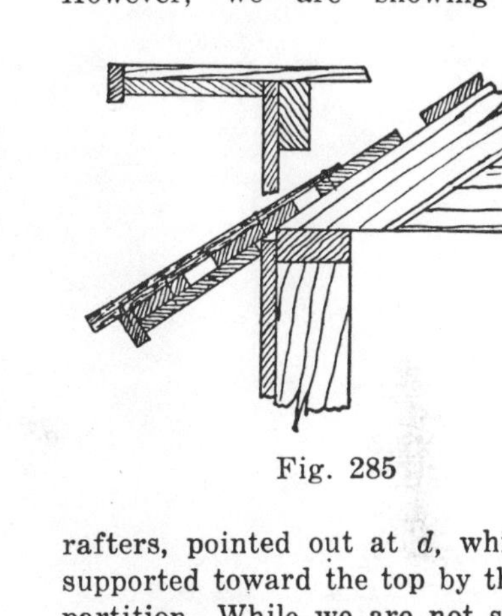

Fig. 285

rafters, pointed out at *d*, which are supported toward the top by the attic partition. While we are not showing any ridgeboards in this roof plan, in most instances it is advisable to use ridgeboards.

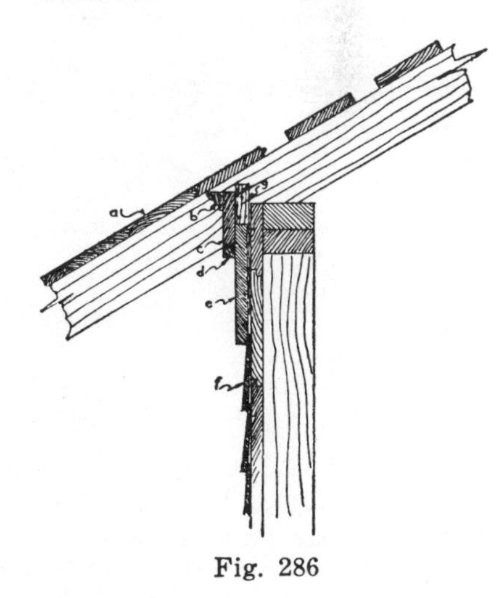

Fig. 286

Tails of Barge Rafters.—Fig. 275 shows a gable of a roof with barge rafters, which are supported by means of four brackets. Fig. 276

shows a simple bevel tail cut on the end of a barge rafter, which is similar to the two forms shown by Fig. 277. Somewhat different cuts for

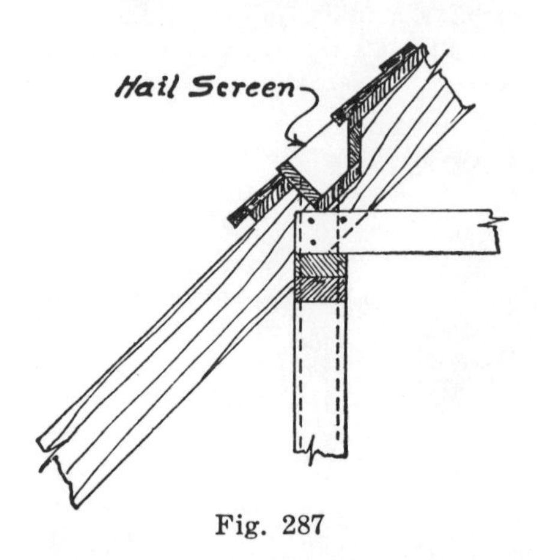

Fig. 287

tails of barge rafters are shown by Fig. 278. A great many modifications of these cuts are possible.

Open House Cornices.—Fig. 279 shows a quite simple and inexpensive cornice. Fig. 280 shows a more sub-

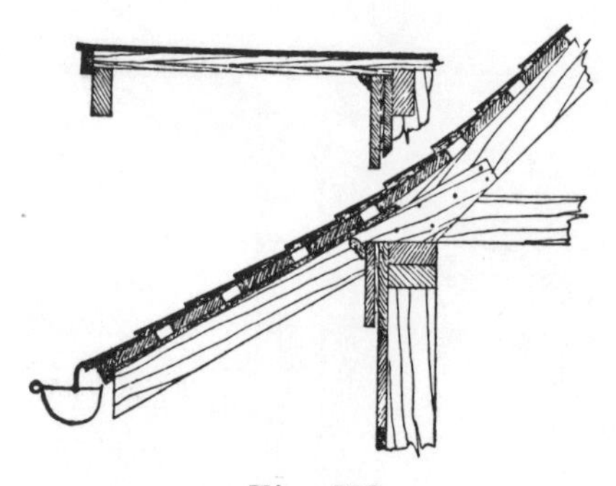

Fig. 288

stantial construction, but still simple. A similar construction, giving a detail of the sheeting boards at the upper left, is shown by Fig. 281. Fig. 282 shows another simple open cornice, giving the rake construction at the top to the left. Fig. 283 shows three gutters in connection with cornices. The upper one is a half-moon gutter. At the center is a Yankee gutter. The one at the bottom is called a crown-mold gutter.

More Open Cornices. — Fig. 284 shows a light cornice, suitable for small buildings. At the upper left is shown the rake construction. A somewhat similar cornice design is shown by Fig. 285. A detail of the rake is shown at the upper left. A cornice construction in part is shown by Fig.

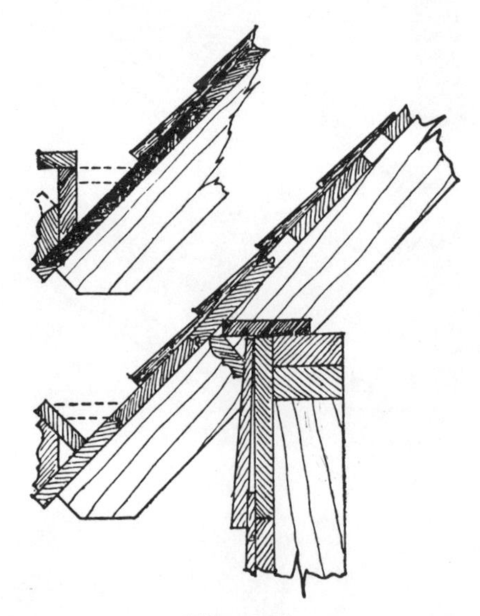

Fig. 289

286. The parts to be noted are: *a*, center matched sheeting; *b*, bed mold; *c*, closer; *d*, a square strip; *e*, frieze, and *f*, siding. Fig. 287 shows a cornice, in part, with a gutter. The gutter has a concealed downspout and is covered with hail screen to keep leaves from lodging in it. An economical sway cornice is shown by Fig. 288. The rake construction is shown to the upper left. Two designs of narrow cornices are shown by Fig. 289, together with Yankee gutters. The dotted lines show the high points of the gutter bottoms.

LESSON 31

THE STEEL SQUARE

Almost everyone knows what a steel square is; that is, he knows that it has a large arm and a smaller one branching off from it at a right angle. The large arm is 2 inches wide and 24 inches long, while the smaller one is only $1\frac{1}{2}$ inches wide

Fig. 290

and 16 inches long. The large arm is called the body, and the small one is called the tongue.

The Steel Square.—Fig. 290 shows a steel square. The intersection of the outside edges of the body and the tongue of a square is called the heel. The face side of a steel square is the side on which the manufacturer's name is stamped. If the body of the square is held in the left hand and the tongue in the right, in a level position, the face side of the square will be up. The side of the square opposite the face side is called the back.

Roof Framing Tables.—Few roofs are framed without the use of the steel square, and while I have never used the tables on the square for any practical purpose, I will explain them in this lesson. The square that I am using as a guide is marked R100C. If the student will lay this square before him with the face up, he will find to the left of the body, top line, these words: "Length of main rafters per foot run," just as shown in Fig. 291. Then directly under the figure 18 he will find 21.63, which means that the length of the main, or common rafter per foot run, with an 18-inch rise, is $21\frac{63}{100}$ inches. To find the length of the rafter, multiply $21\frac{63}{100}$ by the number of feet in the run of the roof. In the same way the second line gives the length per foot run for both the hip and the valley rafters, under the figure (in this case 18) that represents the rise per foot run. But if the rise were only 8 inches per foot run, then you would pass up all fig-

23 22 21 20 19	18	17	16
LENGTH OF MAIN RAFTERS PER FOOT RUN	21 63	20 81	20 00
" HIP OR VALLEY " " " "	24 74	24 02	23 32
DIFFERENCE IN LENGTH OF JACKS 16 INCHES CENTERS	28 84	27 74	26 66
" " " " " 2 FEET "	43 27	41 62	40
SIDE CUT OF JACKS USE THE MARKS ΔΔ ΔΔ	6 11/16	6 15/16	7 3/16
" " HIP OR VALLEY " " " XX XX	8 1/4	8 1/2	8 3/4

22 21 20 19 18 17 16 15 14

Fig. 291

ures excepting those under the figure 8. The figures to be used are always found under the figure representing the rise per foot run. The third line gives the difference in the lengths of the jack rafters, spaced 16 on center; while the fourth line gives the difference in length of jacks for 2-foot spaces. The fifth line reads: "Side cut of jacks, use the marks (inverted V's)." Under the figure 18 will be found 6 11/16. With these figures in mind, pass on to the figure 6, which is shown in Fig. 292. Now count eleven sixteenths on the square and you will find the V-mark, which indicates the graduation mark to be used on the outside edge of the body of the square, with 12 on the tongue. To get the cuts use 12 and 6 11/16 on the square, the arm on which the 12 is used gives the edge bevel for the side cut of the jack rafter for a roof with an 18-inch rise per foot run. The sixth line gives the figures to be used with 12 in order to obtain the side cut for hips and valleys. The graduation marks to be used are indicated by X's. Under the figure 18 on this line, we find 8¼. Now refer to Fig. 292 and you will find the X mark directly under the graduation mark to be used, or 8¼. To get the cut use 12 and 8¼ on the square; the arm on which 12 is used gives the edge bevel for the side cut for hip and valley rafters for a roof having an 18-inch rise per

9	8	7	6	5	4
15 00	14 42	13 83	13 42	13 00	12 65
19 21	18 76	18 36	18 00	17 69	17 44
20	19 23	18 52	17 875	17 33	16 87
30	28 84	27 78	26 83	26	25 30
9 3/8	10	10 3/8	10 3/4	11 1/16	11 3/8
10 5/8	10 7/8	11 1/16	11 5/16	11 1/2	11 11/16

Fig. 292

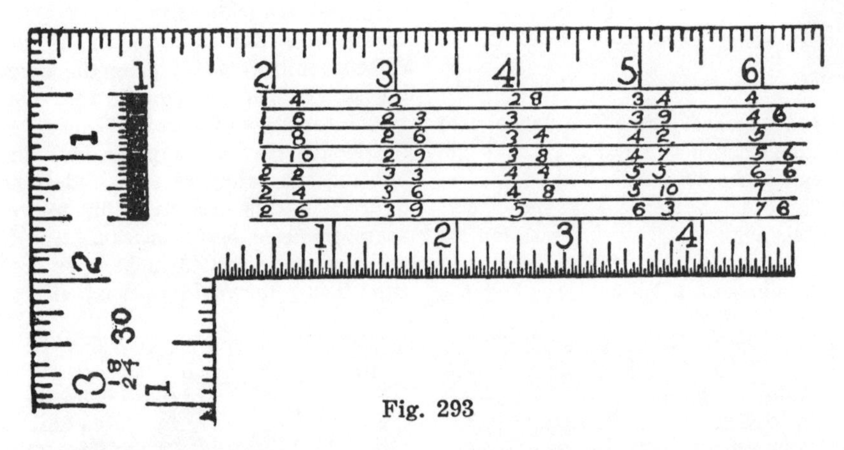

Fig. 293

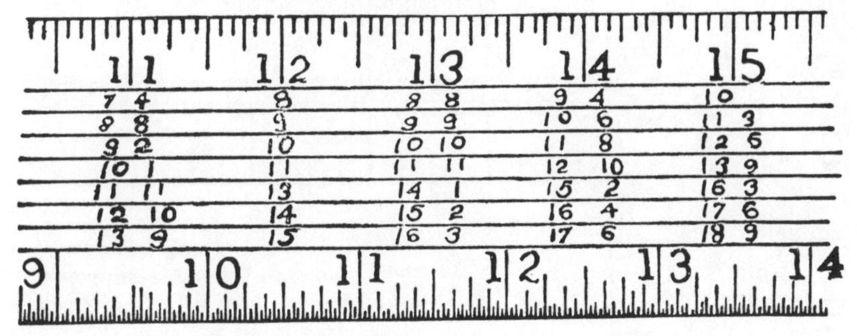

Fig. 294

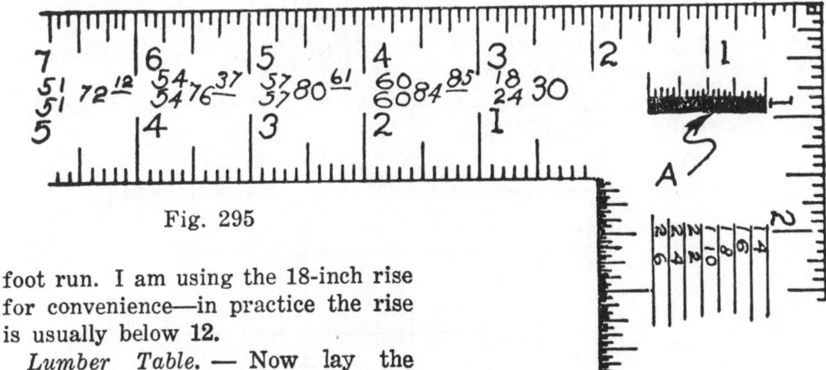

Fig. 295

foot run. I am using the 18-inch rise for convenience—in practice the rise is usually below 12.

Lumber Table. — Now lay the square before you with the back up and you will find what is known as

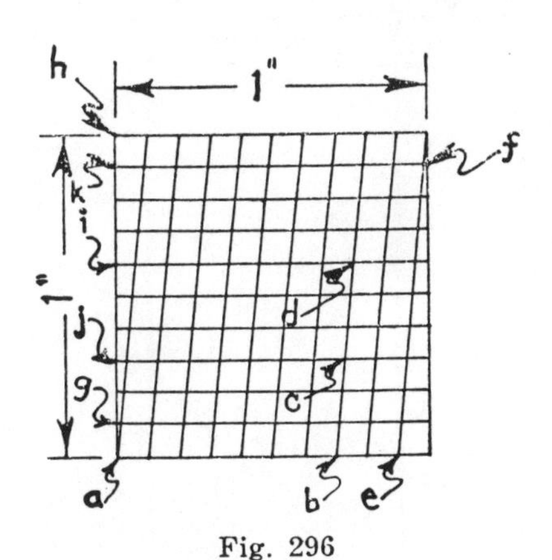

Fig. 296

the lumber table. Figs. 293 and 294 give two sections of the back of the square, showing enough of the lumber table to serve our purpose. The base figure for this table is 12. Now if you will turn to Fig. 294 you will find directly under the base figure, these figures: On the first line, 8; on the second line, 9; on the third line, 10; and so on down to 15. The figure 12 has been omitted. The table can be used in two ways. First, let the figures on the edge of the square represent the length of the board, and the figures under the base figure, the width. Then a board, 12 feet long, 8 inches wide, has 8 feet of lumber in it; if the board is 13 feet long, it has 8 feet, 8 inches of lumber in it; and if 14 feet long, it has 9 feet, 4 inches of lumber in it. These figures are found in Fig. 294 on the part of the square shown and on the upper line. For instance, we have a board 8 inches wide and 3 feet long.

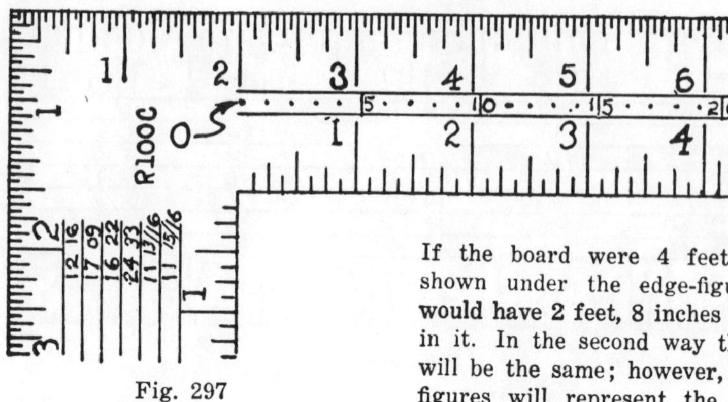

Fig. 297

To get the answer we will turn back to the edge-figure, 3, found in Fig. 293, and under it on the top line we find that it contains 2 feet of lumber.

If the board were 4 feet long, as shown under the edge-figure, 4, it would have 2 feet, 8 inches of lumber in it. In the second way the results will be the same; however, the edge-figures will represent the width of the board and the figures under the base figure, the length. For example, a board 8 feet long, 13 inches wide, has, as we find under the figure 13, 8 feet, 8 inches of lumber in it. If

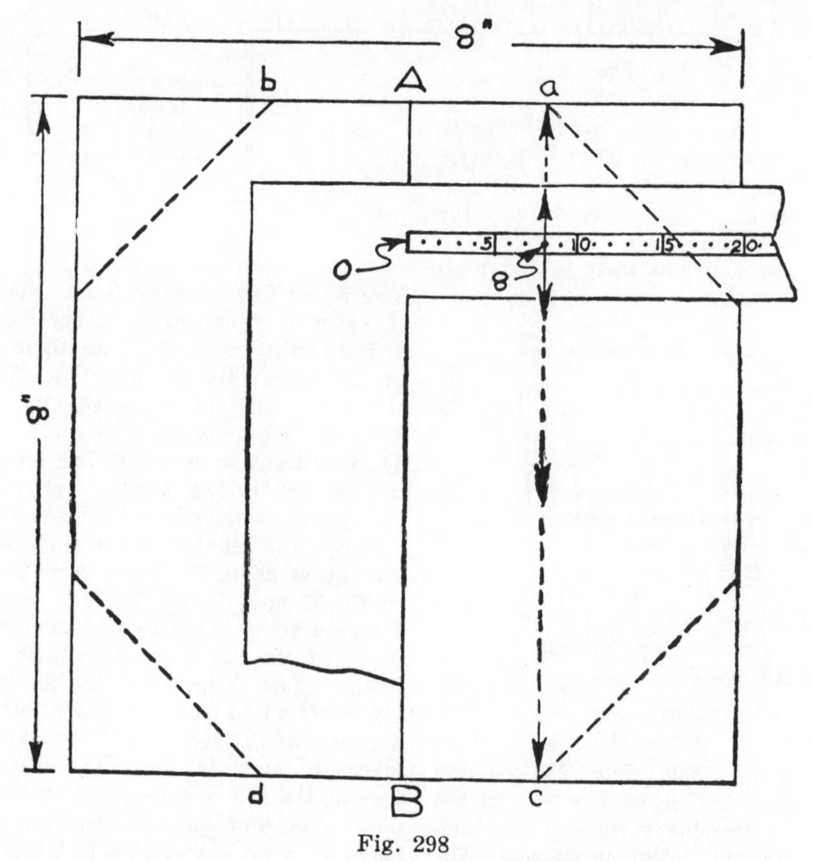

Fig. 298

it were 8 feet long and only 3 inches wide, it would have, as we find under the edge-figure 3, 2 feet of lumber in it. The lumber table is very simple, although it is necessary to understand it thoroughly before it can be used with certainty.

Brace Table.—The brace table is found on the back of the tongue, a section of which is shown by Fig. 295. Under the edge-figure, 3, a little to the right we have the figure 18 has one inch graduated into one hundred parts, in order to make it possible for the workman to obtain the exact length of a brace, or anything else, even to a hundredth part of an inch. This graduated inch is pointed out at *A*, Fig. 295. The method usually used, to obtain the distance of a certain number of hundredth parts of an inch, is by setting the points of a compass exactly at the required number of hundredth parts of an inch

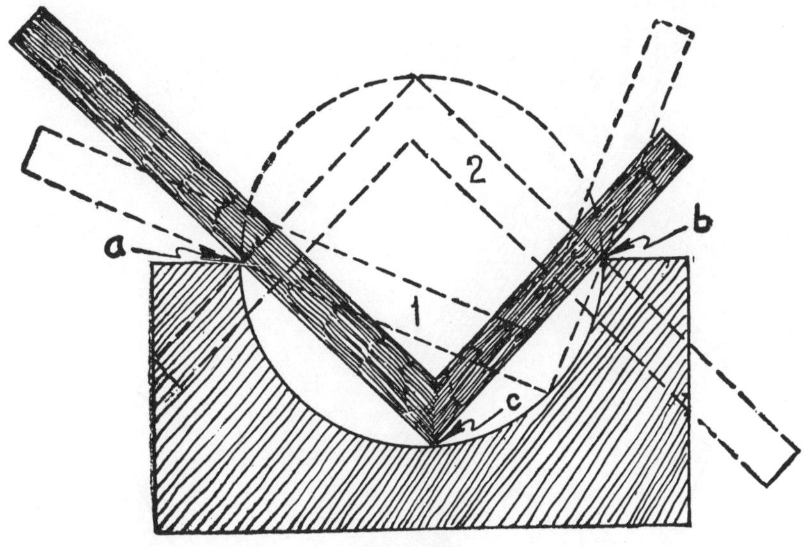

Fig. 299

and below it 24, with the figure 30 to the right. This means that a brace 18 inches from the angle one way and 24 inches from the angle the other way will be 30 inches long; in other words, the diagonal distance between 18 and 24 on a square is 30 inches. Under the edge-figure, 4, we find 60 and 60, with $84\frac{85}{100}$ to the right. A brace put into a square angle, 60 inches each way from the angle would be $84\frac{85}{100}$ inches long. In the same way the lengths of other braces are given in the table.

The square we are using as a guide apart, and adding that distance to the distance measured off with a rule or some other measuring instrument.

Diagonal Scale. — Some squares have what is called a "diagonal scale," like the one shown in Fig. 296. The measuring here is also done with a compass, as in the case just explained, but the count is somewhat different. The diagonal lines as well as the horizontal lines are exactly $\frac{10}{100}$ths of an inch apart. From left to right, as from *a* to *b*, each space counts $\frac{10}{100}$ths of an inch, or in all $\frac{70}{100}$ths of an inch; but from

j to c, 3 spaces up, the distance is $^{73}/_{100}$ths of an inch, and from i to d, 6 spaces up, the distance is $^{76}/_{100}$ths of an inch. Now from a to e the distance is $^{90}/_{100}$ths of an inch, but from k to f, 9 spaces up, the distance is $^{99}/_{100}$ths of an inch. From h to the upper right-hand corner, 10 spaces up, we have $^{100}/_{100}$ths of an inch, or as shown, 1 inch. The space between the two lines at g, 1 space up, is $^{1}/_{100}$th of an inch.

Octagon Table.—The octagon table is found on the center of the face of the tongue, as shown by Fig. 297. The distance here is divided into equal spaces shown by the dots, with every fifth space numbered, as, 5, 10, 15, and so on. In case we were to make an octagon out of a square, 8 inches by 8 inches, as shown by Fig. 298, the first operation would be to draw line *A-B* through the center. Then with the compass take the

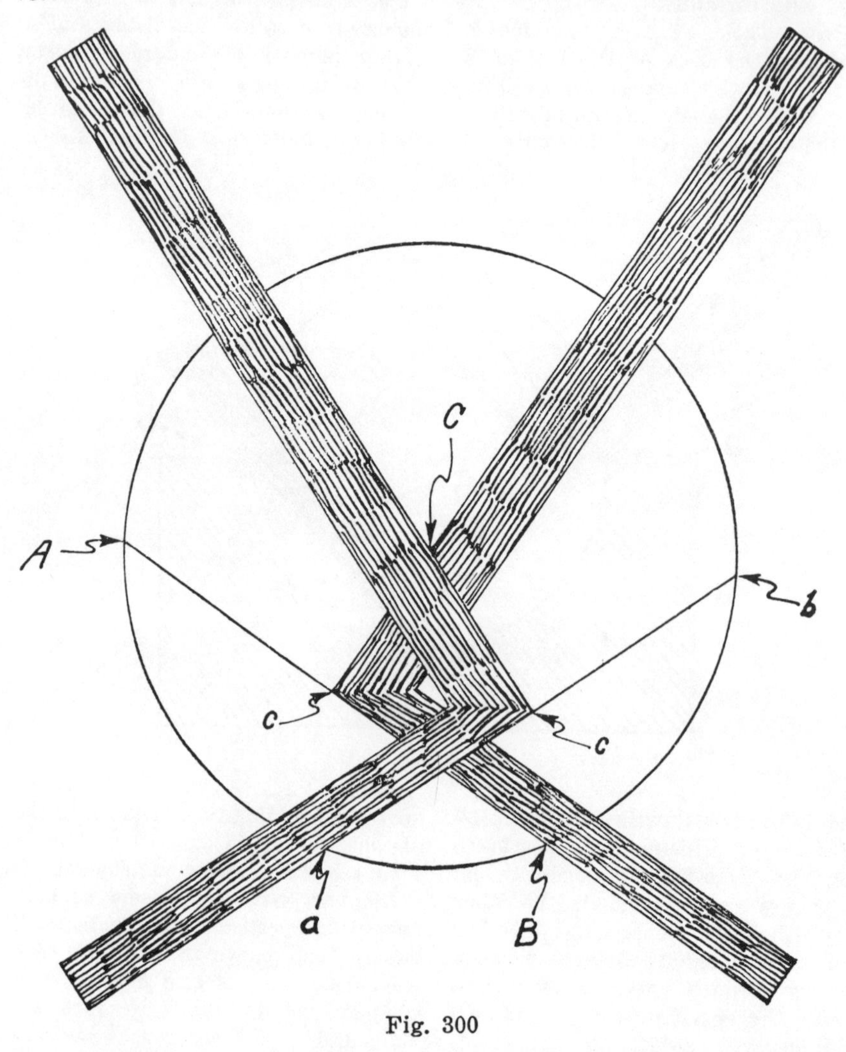

Fig. 300

distance of eight spaces, or the distance between the point indicated at 0, and the point indicated with the figure 8, and lay off *A-a* and *A-b;* also *B-c* and *B-d.* Having these points, it is an easy matter to cut off, as it were, the corners with the 45-degree angle lines, which are represented by dotted lines on the drawing. The arrows in the line between *a* and *c* show another method of obtaining these points. One side marked, the square would have to be reversed to mark the other side. It should be remembered that the width in inches represents the number of spaces to use on the compass; if the width is 8 inches, as shown in Fig. 298, take eight spaces; if the width is 10 inches, use ten spaces; and if 12, twelve. Briefly, use as many spaces with the compass as there are inches in the width of the material.

Graduation Marks.—The R100C square that we are using has the graduation marks spaced one-half inch, one-quarter inch, one-eighth inch, one-tenth inch, one twelfth inch, one-sixteenth inch, one-thirty-second inch, and one-hundredth inch. The one-twelfth inch graduations are the most important, and we would not consider a square complete without having them on one of its outside edges.

Half Rounds.—Fig. 299 shows how the square can be used to test a half round. If the edges are kept in contact with the points indicated at *a* and *b*, the square shown shaded will describe a true half circle with the heel, indicated at *c*. Compare the position of the shaded square with the position of the square marked 1, shown by the dotted lines. The position of the square marked 2 shows how the other half of the circle can be described by keeping the edges in contact with *a* and *b*, and moving the heel from one point to the other as the marking is done.

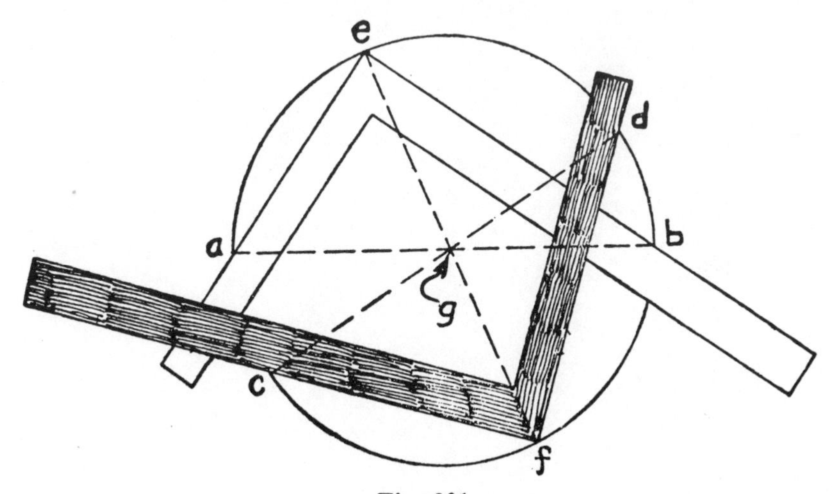

Fig. 301

Methods of Finding Center of a Circle.—Fig. 300 shows how the center of a circle can be found by means of the steel square. Draw two lines somewhat on the order of line *A-B* and line *a-b*, at any convenient position. Establish the center of the two lines, as at *c, c,* and place the steel square in the positions shown. The point at C is the center of the circle.

Another method of finding the center of a circle with the steel square is illustrated by Fig. 301. Place the square in two positions, somewhat as shown, making the heel come in contact with the circle and draw lines along the outside edge of the square. Then join a to b, and c to d, as shown with dotted lines; the point where the two lines cross is the center of the circle. One of the two lines just mentioned can be omitted and a line drawn between e and f, which will also establish the center at the crossing.

Standard Size Square.—The standard size square has a body 24 inches long and a tongue 16 inches long. There are squares that have 18-inch tongues, but they are almost, if not altogether, out of use. Some carpenters like to carry, besides a full-size square, a small square with a 12-inch body and an 8-inch tongue. Such a square has several advantages over the full-size square, especially where space is limited.

LESSON 32

THE SQUARE AND OTHER PROBLEMS

Proving Squares.—Fig. 302 shows a square with the face side up, and the three important parts named, the body, the heel, and the tongue. Fig.

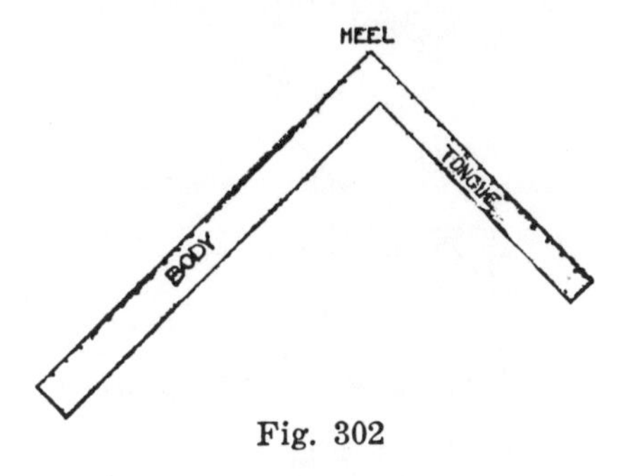

Fig. 302

303 shows how to prove a steel square as to whether or not it is true. Take a board about 12 inches wide with a

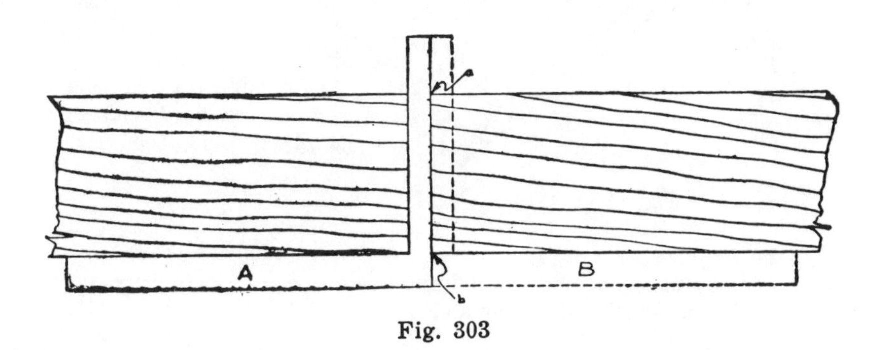

Fig. 303

perfectly straight edge. Then apply the square as shown at *A;* with a sharp pencil draw a line from *a* to *b* along the outside edge of the tongue. Reverse the square to the position indicated at *B*, shown by dotted lines. If the outside edge of the tongue is parallel with the pencil line between *a* and *b*, the square is true. If the test shows that the line and the edge of the tongue do not coincide, say, as shown by Fig. 304, then the square is not true, because the tongue has been sprung outward. On the other hand, if the test brings out the result shown by Fig. 305, then the tongue of your square has been sprung inward and the square is not true.

Circles with the Square. — Stick two nails, as at *a* and *b*, Fig. 306, as far apart as the diameter of the circle you want. Place the square in somewhat the position shown, and with a pencil at the heel mark the circle as the square is moved in the direction of the arrows. When you have one-half of the circle marked, place the square in the position shown by dotted lines in Fig. 307, and proceed to mark the other half of the circle. In case you want a circle that

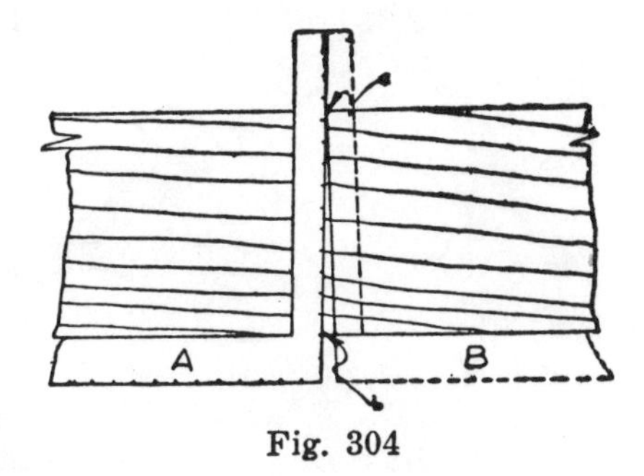

Fig. 304

is larger than what the steel square will make, make a wooden square as large as needed to make the circle.

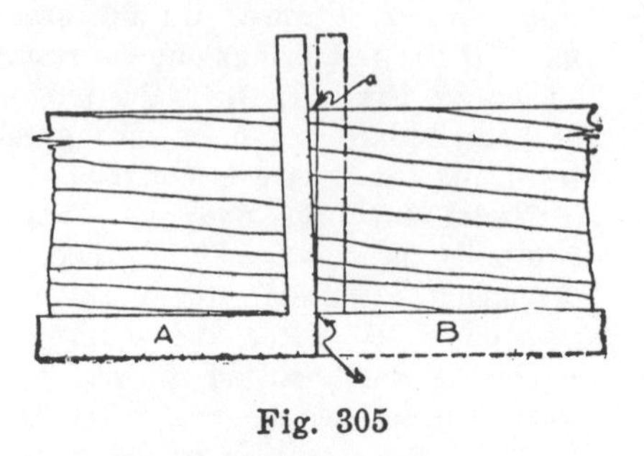

Fig. 305

Fence and Guides.—Fig. 308 shows a square with a wooden fence attached to it. The slots pointed out at *a*, *a*

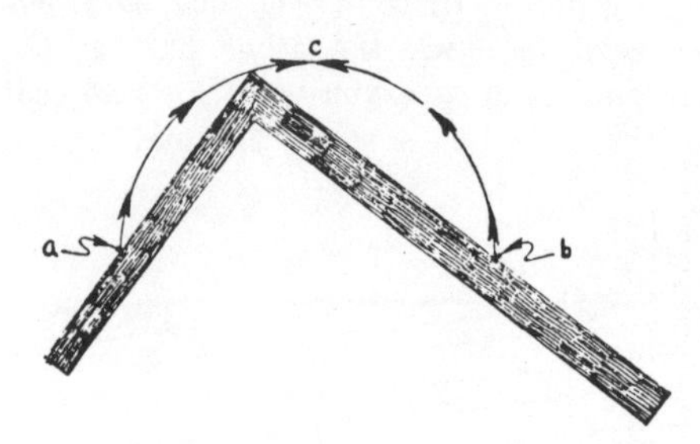

Fig. 306

are used for clamping the fence onto the square by means of the two bolts that are shown both in the upper and

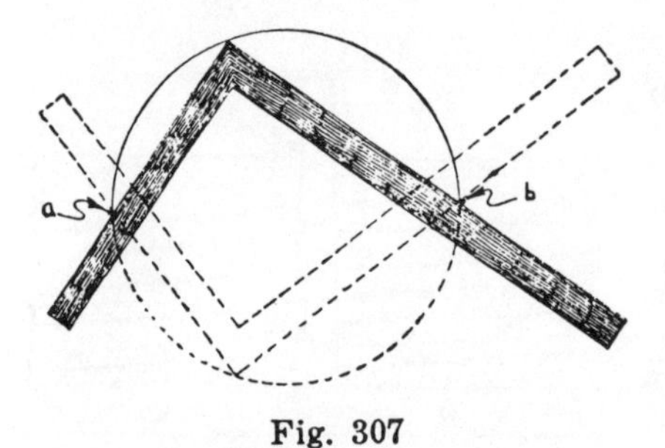

Fig. 307

the lower drawings. The slots pointed out at *b*, *b* receive the square when the fence is attached to it. The wooden fence is not used as much now as it used to be for guiding the square in roof framing. In Fig. 309, at *d* and *e*, are pointed out two metal guides that are much used in framing rafter patterns. At *a* is shown the seat cut of a rafter marked, ready for cutting.

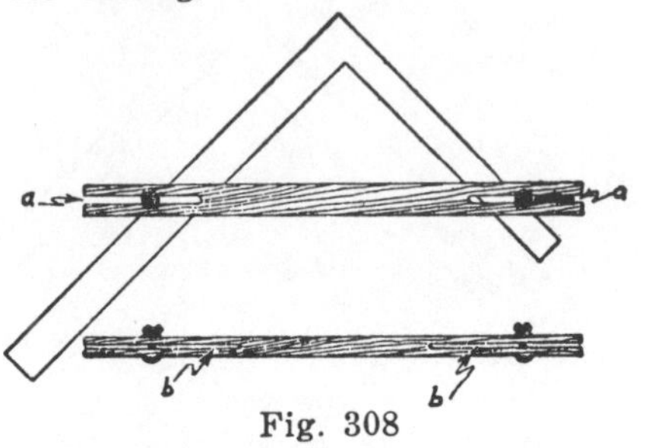

Fig. 308

Square and Compass.—The square and compass are widely different instruments, so far as construction is concerned, and yet used in many cases to obtain the same results. The basic use of the steel square lies in the right angle, but with that instrument a true circle can be described; while on the other hand, the compass is

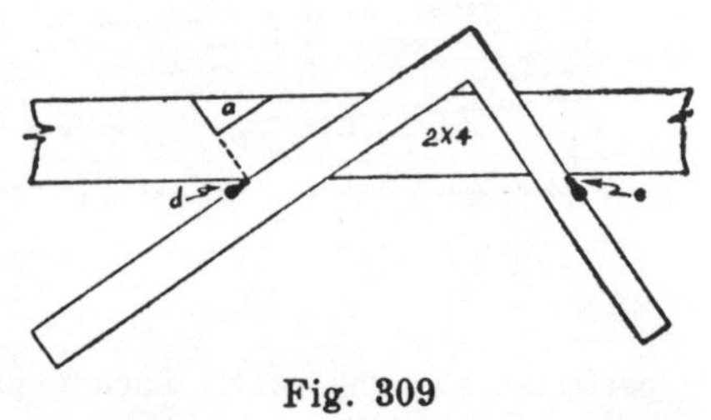

Fig. 309

used primarily for describing circles, but if one has a compass, he can obtain a true right angle. An octagon can be laid out with a square, but just as accurate an octagon can be described with a compass.

Fig. 310 shows how a member cut can be found with a square. The square is shown applied to an obtuse angle, but it does not matter what kind of angle you might have, the principle is the same. We are showing two squares, one shaded and the other not. The unshaded square is placed in such a position that the figure 8 on the tongue intersects with the angle, bringing the blade in a perpendicular position with the base line. The shaded square is so placed that the tongue rests on the inclined line with the figure 8 also intersecting at the angle. Having this, the member cut is obtained by striking a line from the point where the blades of the squares cross, to the point of the angle, or the point where the tongues cross. The dotted line represents this line.

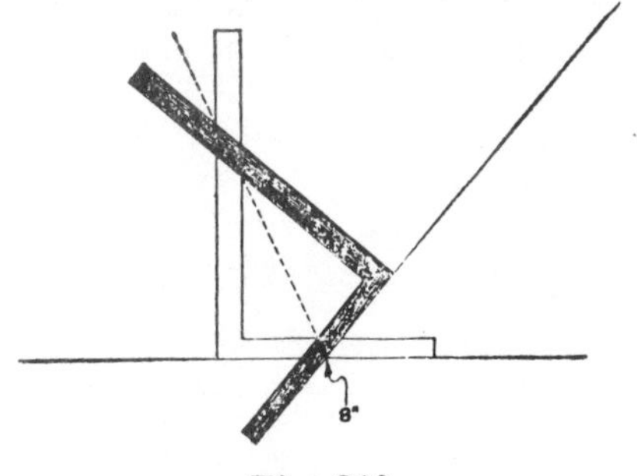

Fig. 310

Fig. 311 shows how to bisect the sharp angle with a square, when the dull angle has been bisected, either with a square or with a compass. In this case the bisecting of the dull angle has been done with a compass, in the manner shown by Fig. 312. The illustration shows two methods. Let us take them one at a time. Place one leg of the compass at point 1, and from this point strike lines 2, 2. Having these points, without changing the compass, strike 3, 3 from points 2, 2. A line drawn through the intersections of 3, 3 will give the miter that will make the molding member. The other method shown in this drawing is just as simple: Strike 4, 4 at any convenient distance from 2, 2 and draw a line from where 4, 4 cross, through point 1, which gives the miter. The same method is used to bisect the sharp angle shown by Fig. 313. Here points *b*, *b* are made from point *a*. Then from points *b*, *b* strike the part-circles *c*, *c*. A line from where these part-circles cross through point *a*, gives the miter.

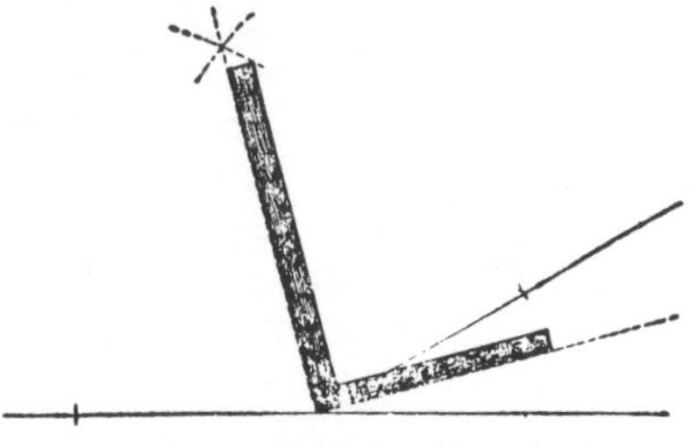

Fig. 311

Curb Cuts. — Fig. 314 shows two ways of obtaining curb cuts, both of which require a drawing somewhat on the order of the diagram.

Proceed by drawing the base line; then draw line *A-B* to the pitch given for the bottom section of the roof, following this by drawing line *B-C* to the pitch given for the upper section

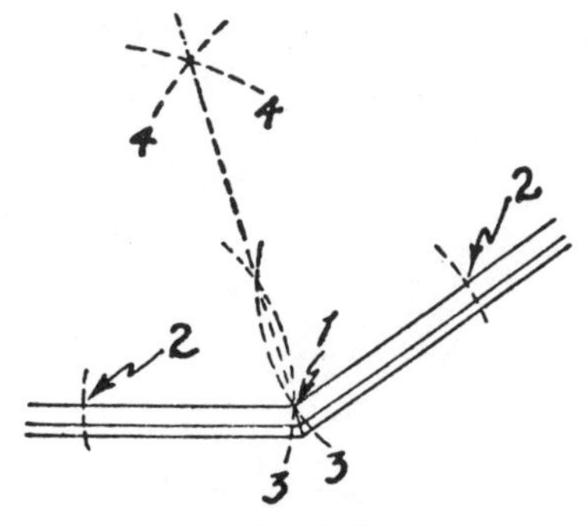

Fig. 312

of the roof. In this instance the bottom section has a three-fourth pitch, and the upper section a one-fourth. Now set the compass at the intersecting point of the roof lines, or point *B*, and at any convenient distance set off *a* and *b*, making them both equal in distance from *B*, as the dotted part-circle will show. These points established, bisect the angle as at *c*, setting the compass at *a* and also at *b*, using a convenient radius. Draw a

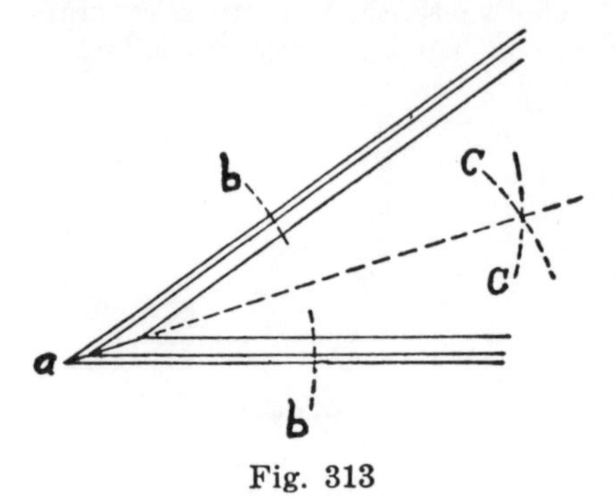

Fig. 313

line from point *B* to point *c*, and you have the curb cut of the rafters.

The same results can be obtained by drawing line 1-2 parallel to line *A-B;* and line 2-3 parallel to line *B-C*, keeping the distance between the lines equal in both instances. By joining

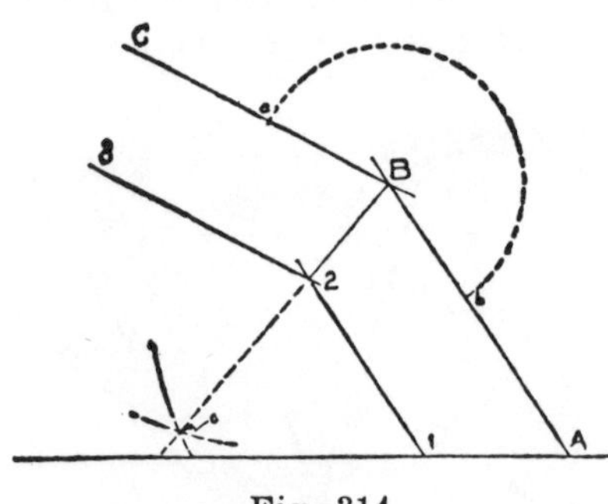

Fig. 314

the two intersecting points with a line, you have the curb cut of the rafters, which is exactly the same as the one obtained in the first solution.

Another method is shown by **Fig. 315.** Here the cuts can all be obtained with the square. The bottom section of the roof, as stated before, has a three-fourth pitch, thus with 12 on the tongue of the square and 18 on the body, both the bottom and the top cuts of the rafter marked *A* can be laid off—the tongue gives both cuts. The upper section being a one-fourth pitch, 12 and 6 will give the cut from *c* to *b*. The cut from *c* to *d* is obtained by taking 18 and 12 and using line *a-c-b* as a measuring line; the arm of the square on which the figure 12 is used will give the cut. Point *c* must be established by the length of the top cut of the rafter marked *A*.

Side Cuts.—A practical way of obtaining the side cut for hip rafters for both regular and irregular roofs is shown by the illustrations. Fig. 316 shows the plan of a roof that is irregular in plan with the hips drawn

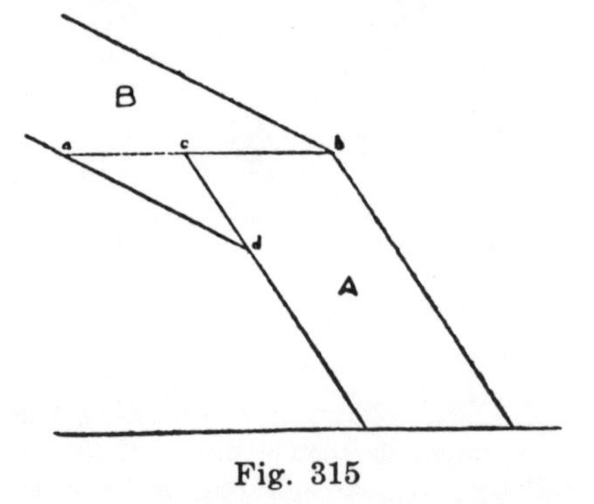

Fig. 315

in. *A*, *B*, and *C* show the three side cuts that are required for the hips of this roof. *D* is the same as *A*. Marking the dotted lines between *a* and *b* of each of the first named corners is the first operation. Fig. 317 gives a perspective view of a short piece of rafter material having the foot cut of the hip rafter. Set this piece on the corner as shown, in line with the rafter line and mark the line between *a* and *b*, as indicated by the dotted line. From point *a* square up to point *c;* and from point *b*, square up to point *d*. Now join *c* and *d*, which gives you the edge bevel for

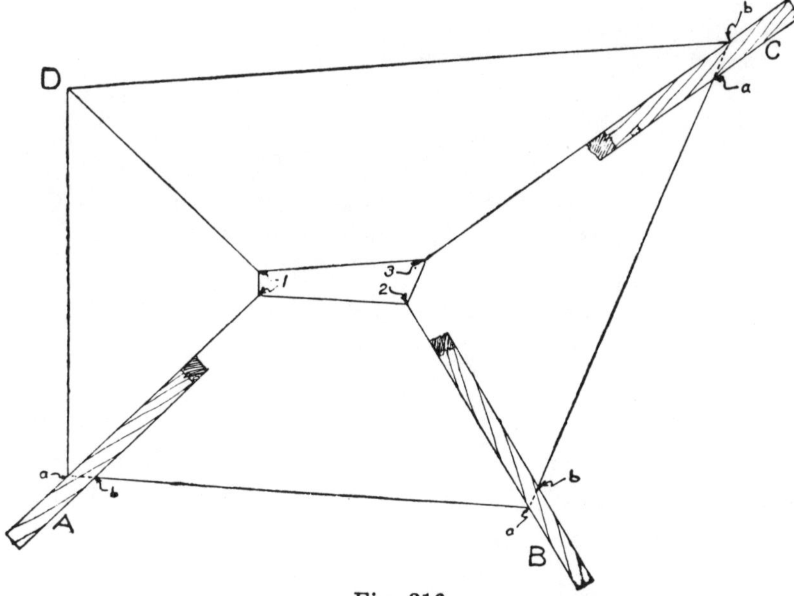

Fig. 316

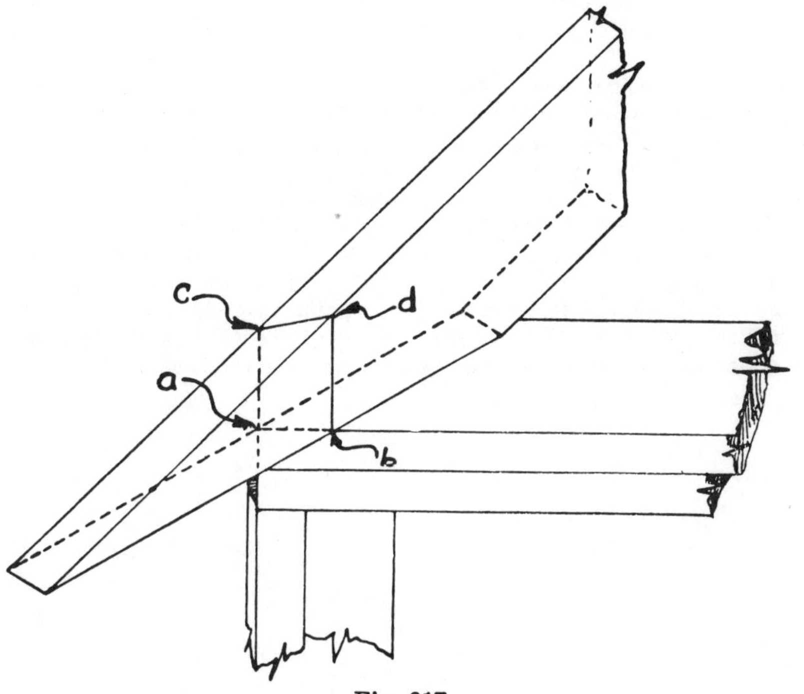

Fig. 317

that corner, say for *A*, Fig. 316. But the bevel is not the same for the corners marked *B* and *C*, so the bevel

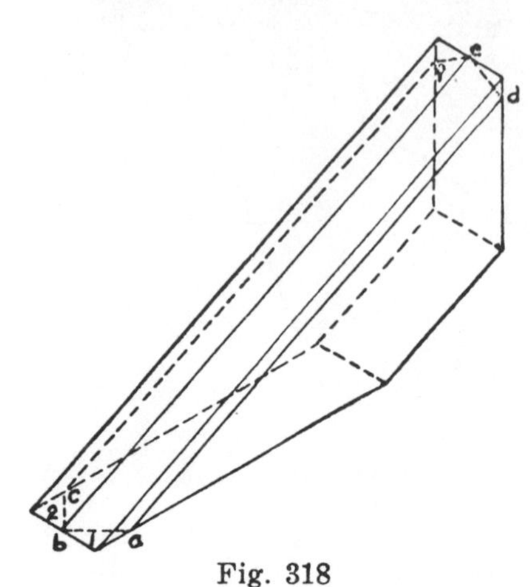

Fig. 318

for each of these corners must be obtained in the same way. These bevels will fit against the deck, 1, 2, and 3, respectively. The two corners marked 1 are the same, but 2 is different and so is 3.

Backing Hip Rafters.—On ordinary residence work hip rafters are not backed, because for all practical purposes the unbacked hip rafter gives as good results and is a little stronger than if it were backed. Nevertheless, there are instances when backing becomes necessary.

Fig. 318 shows the foot-end of a hip rafter marked for the backing. The dotted lines *a-b* and *b-c* show the first operation. These lines are marked at a 45-degree angle and intersect with the center line of the upper edge of the rafter as shown between *b* and *e*. Having the 45-degree angle marks made, make lines *a-d* and *c-f*. Join *f* and *d* with *e*, and proceed to back the rafter by cutting off the two corners marked 1 and 2, at the toe of the rafter. Fig. 319 shows the part of the rafter just mentioned after the backing was done.

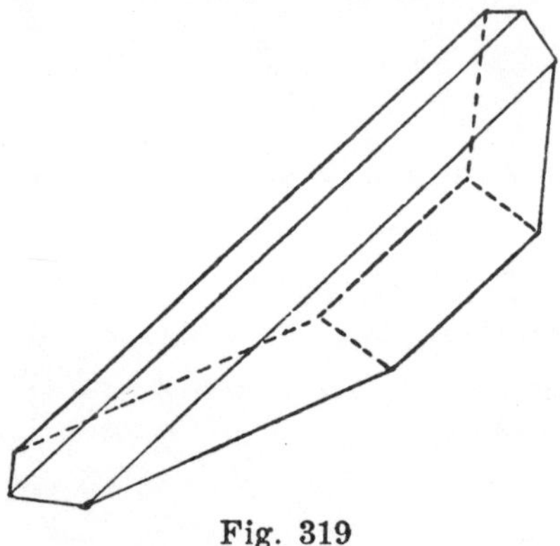

Fig. 319

In cases where the roof is irregular in plan or in pitch, the principle of backing the hip is the same, but the angles of the toe-marks *a-b* and *b-c* must be governed by the relative position the plate lines have to the rafter line; for the lines *a-b* and *b-c* are, in fact, the plate lines. I am assuming that the rafter has no lookout, but should it have, the bevel of the backing is the same.

LESSON 33

RAFTER AND OTHER PROBLEMS

Side Cuts—There is probably no more practical method of obtaining the side cut of jack, valley and hip rafters for any regular pitch roof than the method shown by Figs. 320 and 321. Starting at *A*, Fig. 320, we have a bevel square set to a 45-degree angle, and applied to the seat cut of a rafter in such a manner that by marking along its edge the mark will be on a 45-degree angle. Because jack rafters, speaking of rectangular buildings, always intersect hip or valley rafters on a 45-degree angle (that is, a plan of such rafters would show them intersecting each other on a 45-degree angle), therefore the mark made on the seat of the rafter shown is at the right angle to fit against

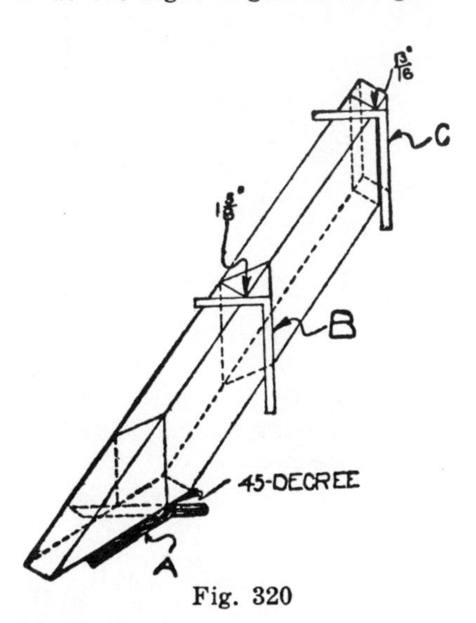

Fig. 320

either a hip or a valley rafter. How to complete the marking is shown by Fig. 321. *A* shows the mark just explained; at either end of this mark we make a plumb mark, as at *a* and *b*. Then we connect the two marks as at *c*—*c* giving the edge bevel for the side cut.

At *B*, Fig. 320, we find the square applied for obtaining the same bevel, but in a different way. Assuming that the rafter is 1⅝ inches thick, place the blade of the square so it will be

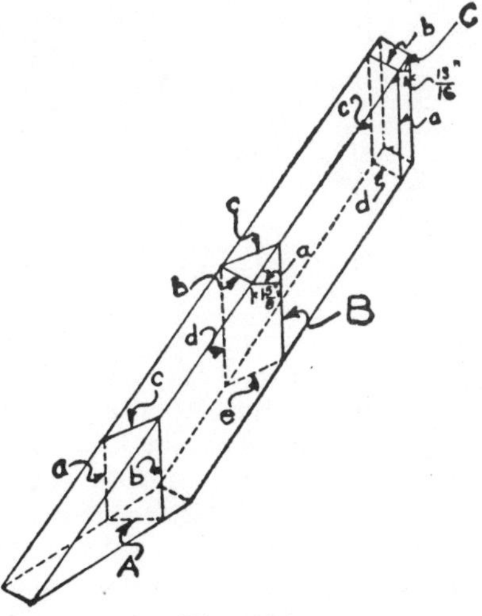

Fig. 321

in perfect alignment with the plumb cut of the rafter you are framing, and then adjust it in such a manner that the 1⅝-inch point on the tongue of the square will intersect with the upper corner of the rafter, as shown. Now turn to Fig. 321, *B*, which is the first mark, or the plumb cut. At *a* is pointed out the 1⅝-inch intersecting line, *b* shows how this point is carried square across the upper edge of the rafter. Now the line shown at *c* can be made, which gives the edge bevel of the side cut, or the same bevel we obtained at *A*. The dotted lines shown at *d* and *e* show the plumb cut of the other side of the rafter and the bevel at the bottom edge, respectively. At

C, Fig. 320, the square is applied for obtaining the amount to be cut off a common rafter to allow for a $1\frac{5}{8}$-inch ridge board. Half of $1\frac{5}{8}$ inches equals $\frac{13}{16}$ of an inch, therefore the square, being in perfect alignment with the plumb cut, must be adjusted

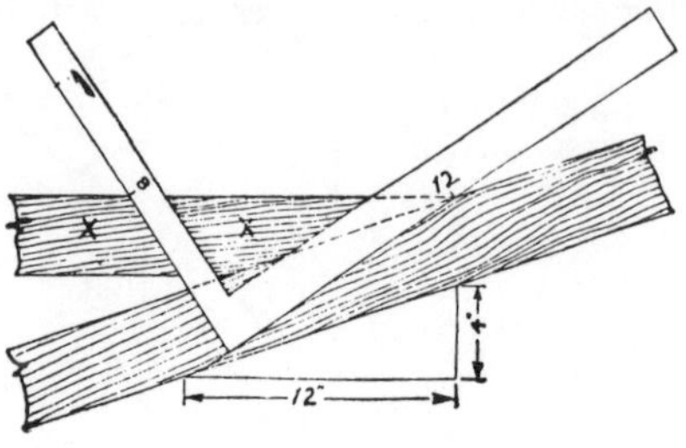

Fig. 322

so as to make the $\frac{13}{16}$ of an inch point intersect with the upper corner of the rafter. Having done this, we carry the point just obtained across the upper edge of the rafter, as at *b*, Fig. 321, *C*. Intersecting with this line we mark the plumb cut, as at *a*. The dotted lines shown at *c* and *d* show, respectively, the cut on the other side and the cut at the bottom.

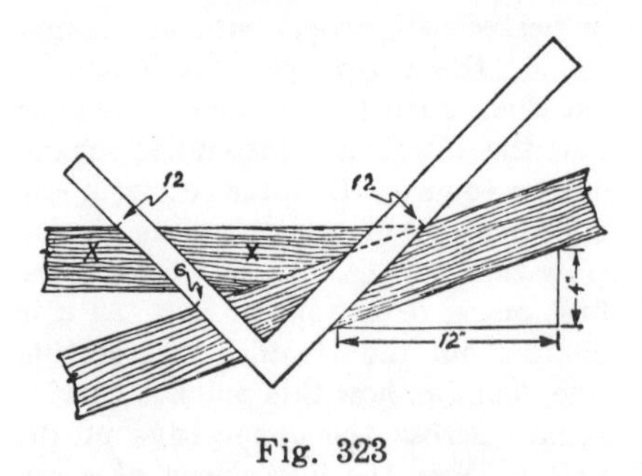

Fig. 323

The principle here is the same as the principle involved at *B*, excepting that this is a plain plumb cut, while at *B* we are showing how to obtain the edge bevel for a side cut.

Fly-Rafter Cut.—Let us assume that we have a main roof with a one-third pitch, to which we are to join a dormer roof with a one-sixth pitch, or a 12-and-4 pitch roof. Fig. 322 shows how to obtain the cut. Take a short piece of material and give it a 12-and-4 cut and tack it to the rafter material in the manner shown by the piece marked *X*, *X*. This, it will be seen, will put the upper edge of the piece in a horizontal position, if the rafter material were in the position of a rafter in place. To obtain the cut, take the figures for a one-third pitch, 12 on the body of the square and 8 on the tongue, and apply the square as shown in the drawing. Where the edge of the tongue intersects with the upper edge of the

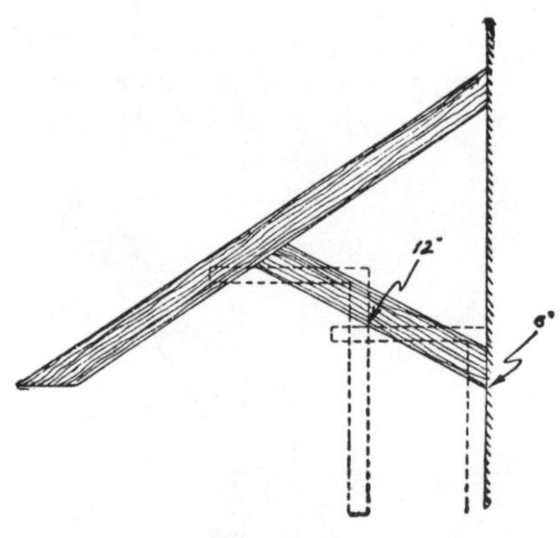

Fig. 324

rafter material is the point to use with 12 to mark the cut. The blade will give the cut.

It is easy to determine which part of the square gives the cut: If the main roof is higher than a one-half pitch, then the smaller of the figures gives the cut, but if the pitch is lower than a one-half pitch, then the larger of the figures used on the square will give the cut, as shown by Fig. 322.

Fig. 323 shows how to get the cut when the same kind of dormer with the same pitch joins a one-half pitch main roof. Here the short piece, marked *X*, *X*, is cut the same as in the other instance, but the application of the square is made to conform with the figures used in framing a one-half pitch roof, or 12 and 12. In this case, as we are showing, 12 and 6 are the figures to use. The body of the square gives the cut. A little study of the two drawings will clarify the problem.

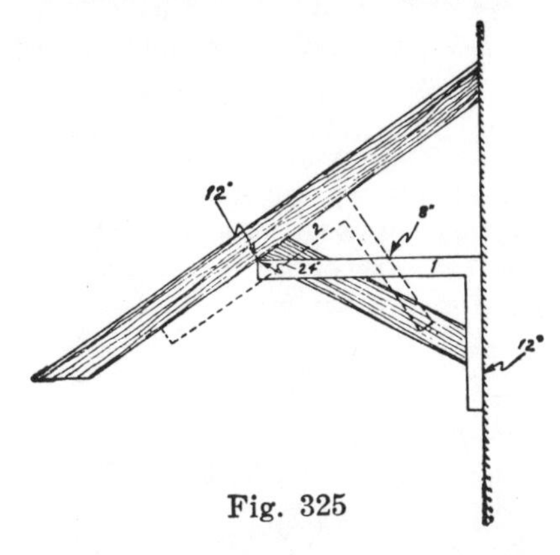

Fig. 325

Cuts of Braces For Hoods.—Fig. 324 shows how to obtain the length of a brace for a hood by stepping it off with a steel square. The figures to be used on the square are 12 and

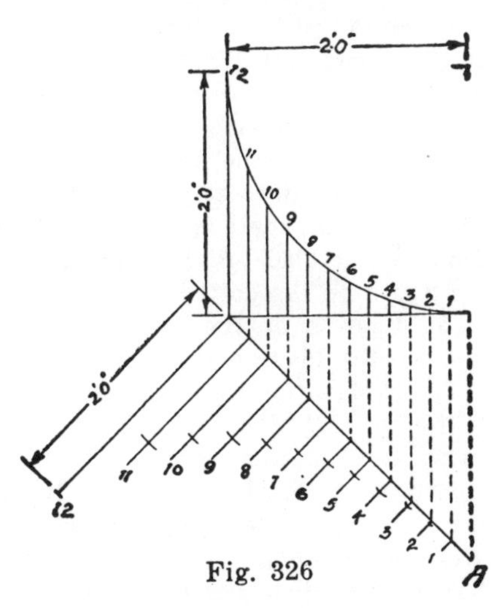

Fig. 326

6, which also give the cut of the brace where it joins the building.

Fig. 325 shows how to get the cut of the brace where it joins the hood rafter. Here square No. 1 is shown applied to the brace in a double-step position, using 24 on the body of the square and 12 on the tongue. The dotted outline of a square, numbered 2, shows how to get the cut by using 12 and 8, the figures used for framing the hood rafter, which has a third pitch.

The principles presented here will work for any pitch, when the figures used on the square conform with the pitch of the rafter and the angle of the brace, respectively.

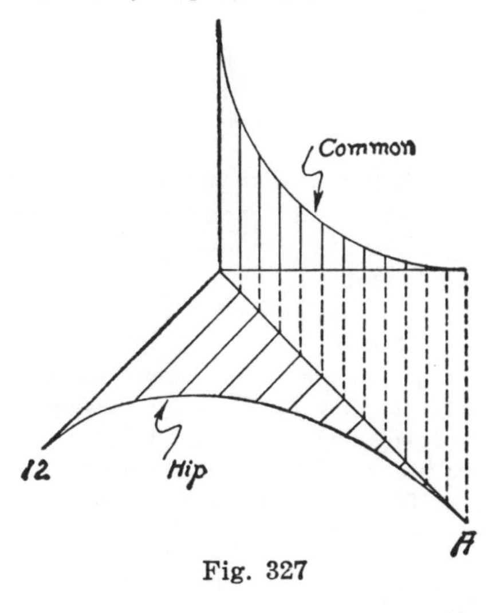

Fig. 327

Obtaining Sways for Hip Rafters.—Fig. 326 is a diagram of a common rafter, the run of which we have divided into 12 equal parts. (Any number of equal parts can be used, the more the better.) These points we have raised perpendicular to the base line, until they intersected with the common-rafter curve. At the bottom of the diagram we are showing the development of the hip curvature. First, we draw the base line of the hip on a 45-degree angle and drop the division points of the common rafter until they strike this line, as indicated by the dotted lines. At each of these intersecting points we draw a line perpendicular to the base of the hip, as shown on the diagram. Then, with a compass we transfer the distance between the base line and curved line of each perpendicular line of the common rafter, to the respective perpendicular line of the hip. In other words, we mark the perpendicular lines of the hip as long as the

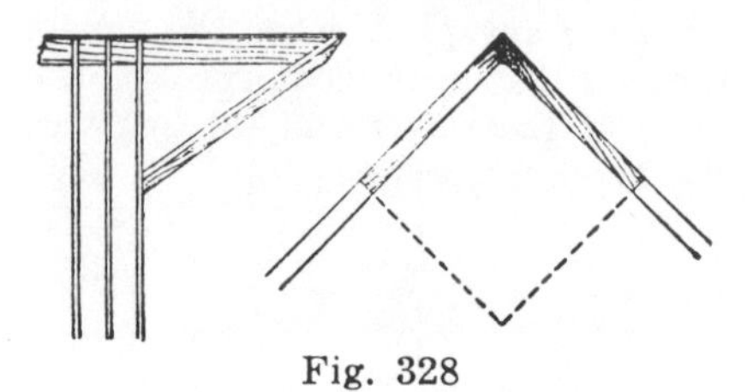

Fig. 328

respective perpendicular lines of the common rafter. Now draw a line from point *A*, crossing these marks, and stop at 12. The curvature between *A* and 12 is the curvature of the hip rafter.

Fig. 327 shows the same diagram with the hip rafter fully developed. You will notice that the curvature of the hip is not a true circle curve, and therefore can not be obtained with a radius.

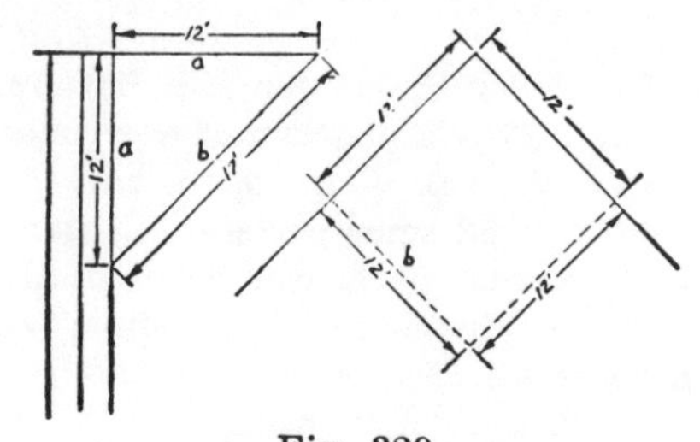

Fig. 329

Hayfork Hood.—Framing the rafters (if they can be called rafters) of a hayfork hood on a barn roof often becomes a problem. If the rafters are to be set plumb, speaking with reference to the sides of the timbers, the cuts are obtained exactly as you would obtain the cuts for hip and valley rafters. In that event the rafters would need backing in order to fit the sheeting. But when the rafters are to be set so the sides will be at a right angle with the sheeting, then you have another problem.

Fig. 328 shows the problem. To the left is a side view and to the right a face view. (All unnecessary lines have been omitted.) Fig. 329 shows a diagram of the same problem. The figures we are using are exaggerations—12 and 12 and 17 are figures that are easily understood, and therefore we are using them.

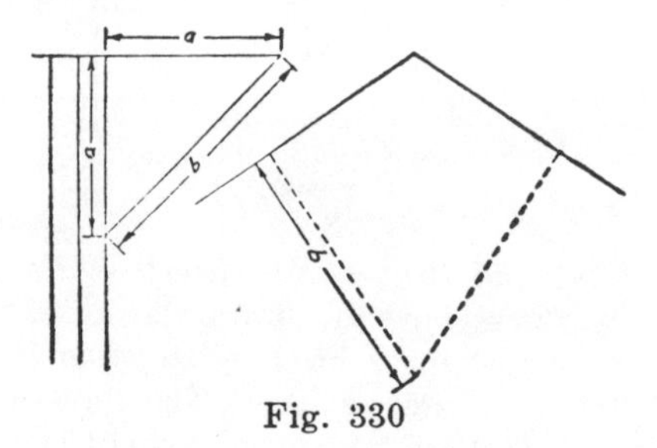

Fig. 330

To obtain the bevel for the edge of the rafter, use the figures on the square shown on the line *a* and *a*. The bottom cut of the side of the rafter is square across, while the top cut on the side is obtained by taking the figure shown on line *b* of the left diagram on the blade of the square, and the figure on the dotted line *b* of the right diagram on the tongue of the square—the blade gives the cut.

The explanation we have just given is for a half-pitch roof. Fig. 330 shows two diagrams drawn with reference to a third-pitch roof. Here the principle is the same, but the figures on the dotted line *b* will be different. The distances of the lines *a* and *a* taken on the square will, respectively, give the cuts for the edge of the rafters. The bottom cut for the side is square across. For the top cut on the side of the timber, take the distance of line *b* of the left diagram and the distance of the dotted line *b* of the right diagram—the former gives the cut.

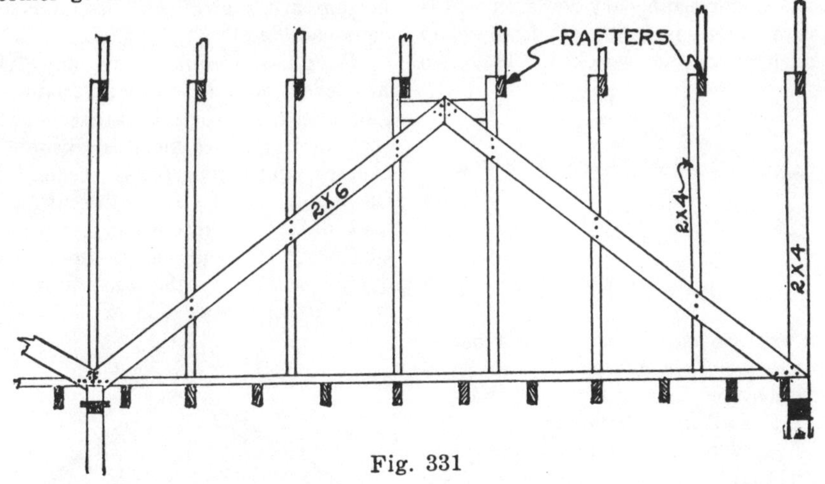

Fig. 331

Trussing a Roof.—Trussing a side wall of an attic that comes over a room, somewhat on the order shown by Fig. 331, is illustrated by this figure. The rafters for the truss should be selected on the basis of the amount of weight they must carry. In this instance we are showing 2x6's. To install them, first see to it that the ceiling joists are lining perfectly, then put the truss in place. There are two ways that this can be done. The rafters of the truss can be spiked to the studding, or, the studding can be notched to receive the rafters, which again must be spiked. The latter method is the most substantial.

LESSON 34

VARIOUS ROOF PROBLEMS

Comb Problems.—Fig. 332 shows a comb construction for the cheapest kind of roof. Such a construction is suitable for temporary roofs, in which the material is to be salvaged when the building is wrecked. The two

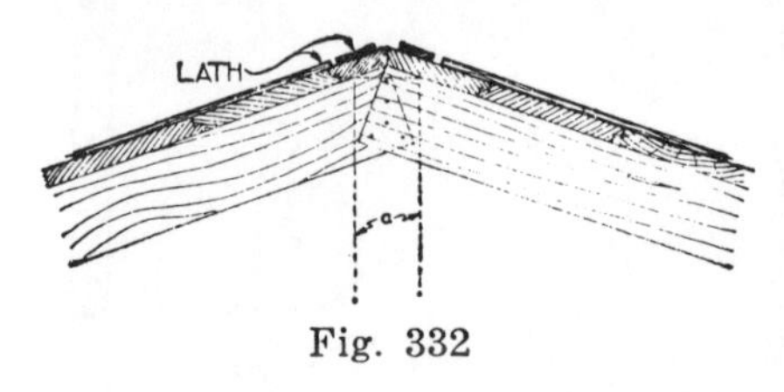

Fig. 332

rafters are merely lapped and nailed together. This leaves the material full length. The collar beam hanger is indicated by dotted lines.

Fig. 333 shows a comb construction that is suitable for a roof on which corrugated sheet iron is used to carry off the water. The upper drawing points out the rafters and the headers, while the bottom drawing gives a cross section of the plate, foot joint, comb construction, and one header to the right.

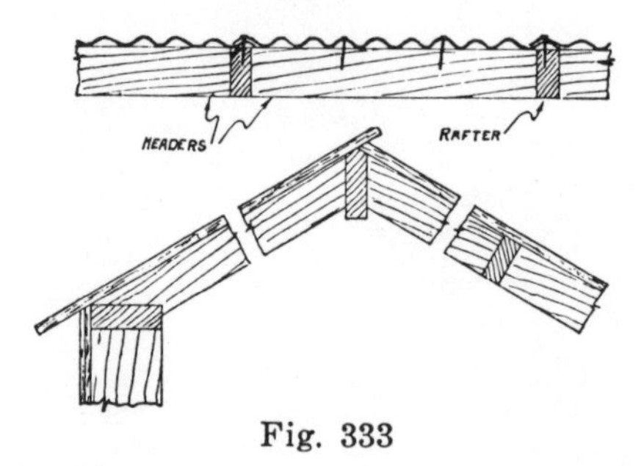

Fig. 333

Fig. 334 shows a comb construction without a ridge board. A metal ridge roll is shown at the top, giving its relationship to the last courses of shingles with figures. Fig. 335 shows the same kind of construction. At the top is shown a metal ridge roll as compared with a comb-board ridge finish, shown at the center. At the bottom are shown three comb board constructions.

Shingling Gauge.—Any one who has ever shingled valleys, using a gauge on his shingling hatchet, knows that it is quite difficult to keep the courses running perfectly straight at the valley. There are different ways that different shinglers use to solve this problem. Some use a line at frequent intervals to reestablish a straight line and also to keep the

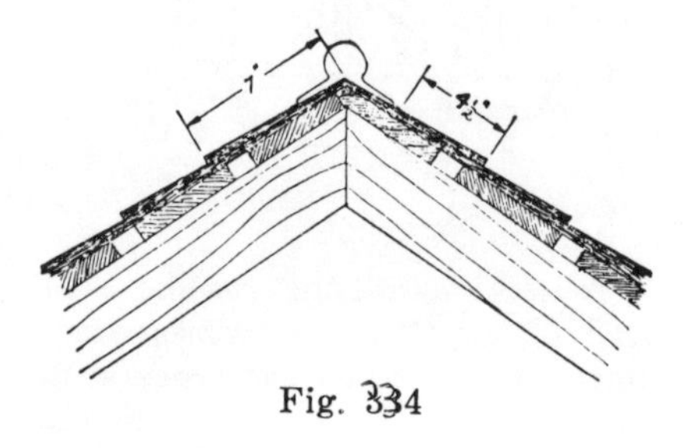

Fig. 334

courses running parallel with each other. Others use a straightedge and mark the courses a short distance from the valley, and still others use a shingling gauge similar to what we are showing by Fig. 336.

The gauge is made of wood and thumb bolts. A 1x3 straightedge, about 6 feet long, with a slot cut about at the center for adjusting the gauge, a 1x2 guide contacting the valley shingles and a 1x2 brace held together with thumb bolts as shown constitute the layout. This gauge can be used over and over on different valleys, because it is adjustable. In cases where such a gauge is needed for one job only, the thumb bolts can be omitted and the joints fastened together with nails.

It is not necessary to use the gauge for starting every course, but it should be used frequently in order to keep the courses running straight and parallel.

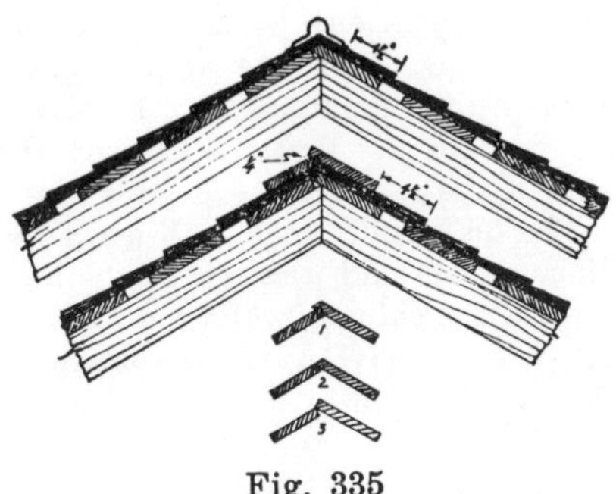

Fig. 335

Marking Cuts for Sprung Moldings.—Fig. 337, to the left, shows a cross section of a large molding in place. The part of the back we are dealing with here is shown to the right, shaded—we are looking straight at it, as indicated by the large arrow to the

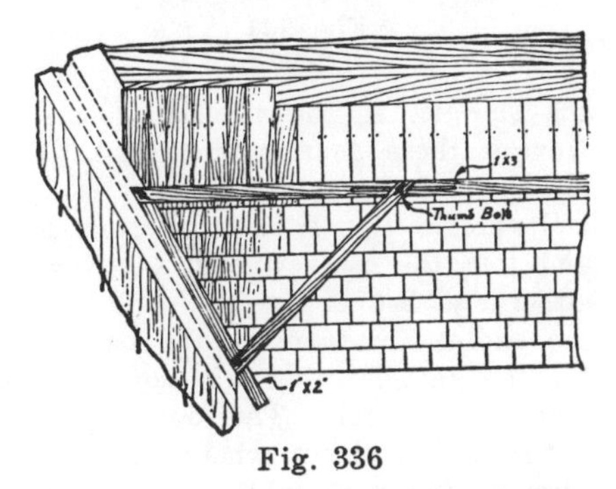

Fig. 336

left, which points to the bottom of the cross section. From the starting point we make the first mark, as between *a* and *a*, which is a square mark. Now turn to Fig. 338, where we show a cross section to the upper left, and a drawing of the back, sloping down toward the right. Here again we are looking straight at the back, as indicated by the large arrow to the left of the cross section. The mark that we make here is between *a* and *a*, and the way we arrive at it is by squaring across the back (the shaded part) as shown by the dotted line. Then we measure to the right the distance of the spring of the molding, as shown by the cross section, which is 3 inches. This gives the second point, and we mark from *a* to *a*. (This marking can also be done with the steel square, by

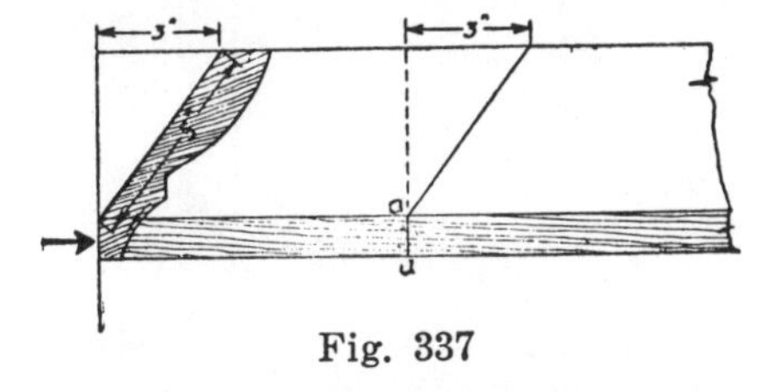

Fig. 337

taking 3 on the tongue of the square and 5 on the body, the tongue giving the cut. Those figures are given with the cross section at the upper left. The cut is the same as a sheeting or plancher cut for a hip roof.)

Fig. 339, to the left, again shows the cross section of the molding, and

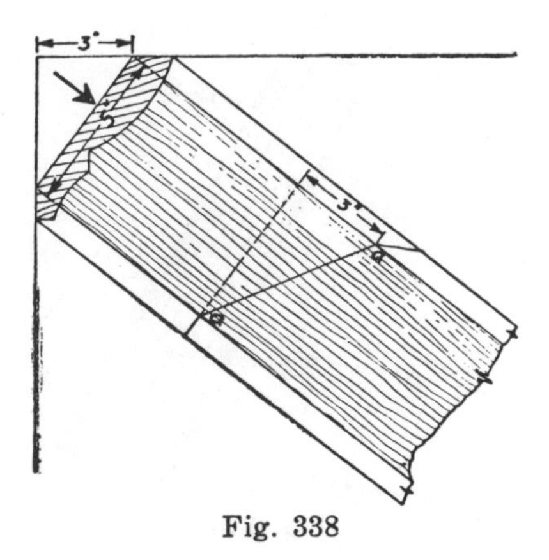

Fig. 338

the arrow shows the direction we are looking at the back. The part below the cross section that is shaded is what we are dealing with here. The mark that we want is between *a* and *a*, which is a true miter, as indicated to the right at *a*, where we show a square applied, using the figures 12 and 12.

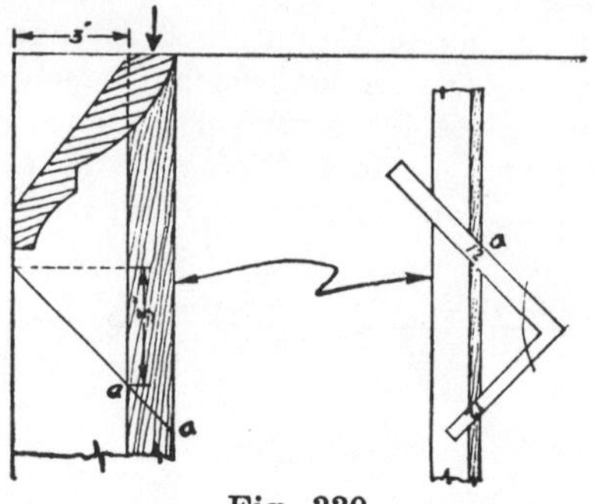

Fig. 339

With the back of the molding marked in this way, take a well-sharpened fine saw and cut the molding from the back so as to cut away

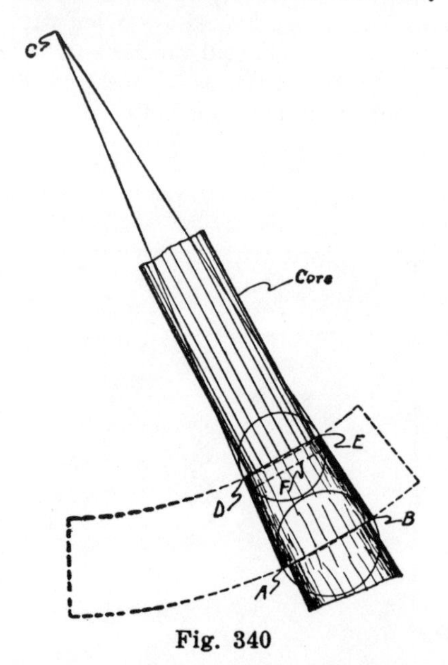

Fig. 340

the three marks. If this is carefully done you will have a true miter cut.

Rolled-Edge Shingles.—Fig. 340, a sort of diagram, shows a part of the core for the roll of a rolled-edge cornice, with two of the shingles in place. To make the butt edge of the shingles run straight around the core of the roll, it will be necessary to cut the shingles to a circular curve, on the order shown by dotted lines on the drawing. What we want to know is the radius for marking the pattern for the shingles that are to cover the core of the roll.

The distance from *A* to *B* gives the diameter through the roll at the butt edge of the shingle, while the distance between *D* and *E* gives the diameter through the roll at the other edge of

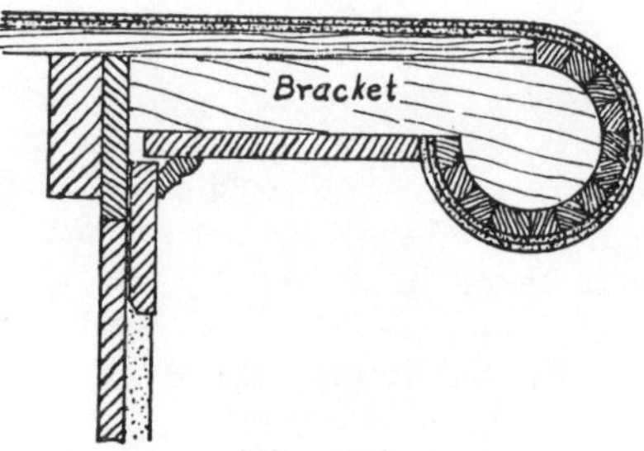

Fig. 341

the shingle, as the circles indicate. Having these four points, strike a line from *B* through *E* and on to *C*, and another line from *A* through *D* and join the other line at *C*. Now, either line *B-C* or line *A-C* will give the radius for marking the pattern for the shingles that will cover the core of the roll. The dotted line shown at *F*, shows where the butt edge of the next shingle will come. The lap of the shingles should not be so wide that it will cause open joints where the shingles lap.

Rolled Edge.—There are many designs for rolled edges for roofs, the shape of which is usually formed by the core. We are showing by Fig. 341 a cross section of a cornice with a rolled edge. This design is simple, and can be modified in different ways, in order to suit the taste of the owner. The bracket is made of 2-inch stuff, the circular part of which is covered with narrow strips running parallel to the rake of the roof. If the rolled part is small the strips will have to be narrow, but if it is rather large, wider strips will serve the purpose. The bracket is nailed to the projecting sheeting. The round end of the bracket is the form for the core.

Fig. 342 shows a face view of a part of a rolled edge on a gable with the shingles in place. To the left we

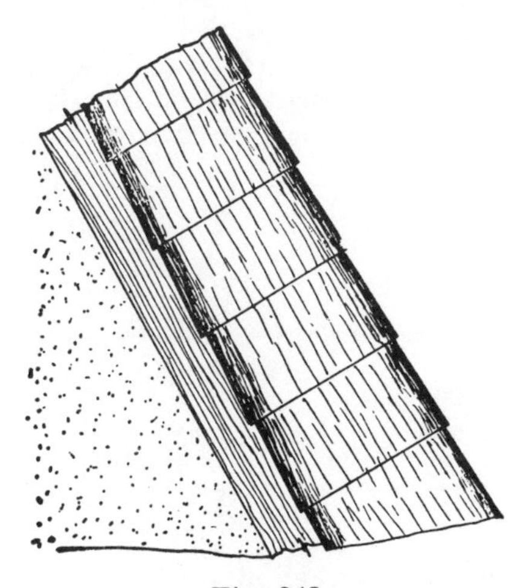

Fig. 342

show part of the frieze and stucco.

Fig. 343 is a practical way to mark the pattern for the shingles that cover the core. At *A* we show a shingle fastened temporarily flat against the core, over which the strip *B* is fastened, making a guide for the scribers with which the dotted line, *a-b*, is made on the pattern, marked *C*. It will be noticed that the shingle marked *C* does not run straight around the roll. This is caused by the butt edge lapping over the shingle marked *A*. When the pattern, *C*, is cut to the dotted line, *a-b*, it will show a curve when it is laid on a flat surface, but when it is fastened on the core the butt edge will run straight around the core. After the butt end is cut to the right curvature, it should be cut to an even width, so as to give the same lap around the core. This would give you a pattern something like what is shown by Fig. 344.

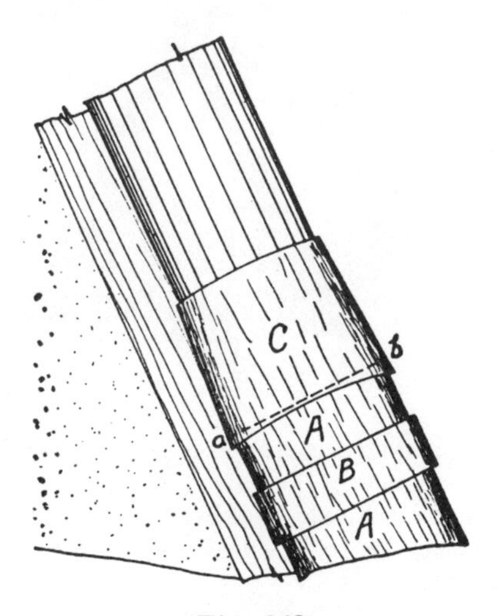

Fig. 343

Shingles for Cone Roofs.—Cutting shingles, especially asphalt shingles, for cone roofs is a simple matter. Such shingles can be cut from slated roll roofing. The different radii for

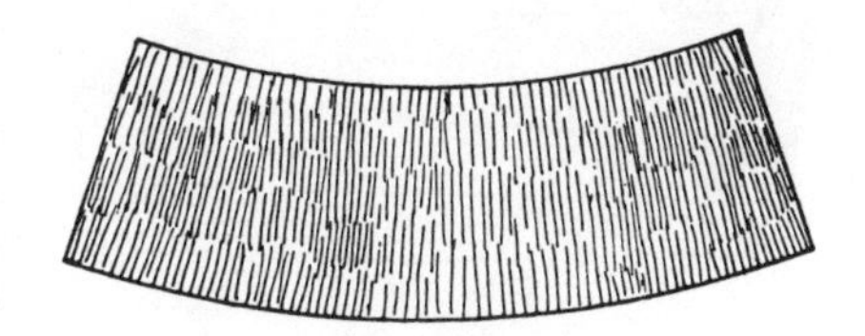

Fig. 344

describing the shingles are easily obtained.

Fig. 345 shows an elevation of a cone roof. The radius for marking the shingles for the first course is the distance between point *A* and the eave at *B*. Such a shingle is shown cut to shape between 1 and *B*. The second course, numbered 2, would have a radius the width of the exposure to the weather shorter than the radius for the first course. The third course would have a radius the width of a course shorter than the radius for

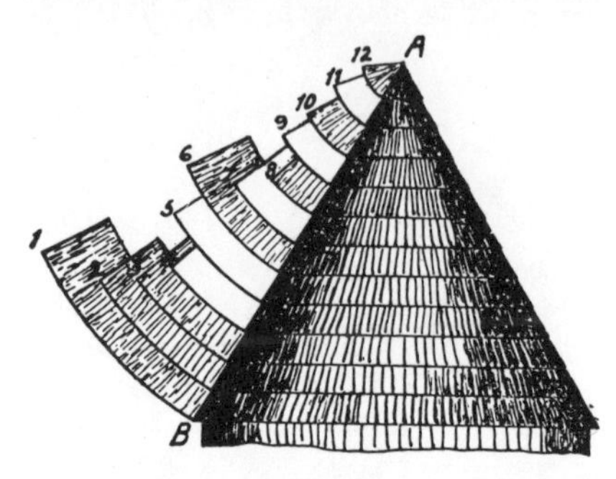

Fig. 345

the second course, and so on, each radius being the width of a course shorter than the radius for the course before, up to the last course, which is numbered 12.

Fig. 346 shows a roll of roofing partly unrolled with a radius pole in place for marking the shingle. To the right the marked shingle is shown shaded. Fig. 347 shows the same roll of roofing, but here the shingles are cut in longer strips. The pivot of the

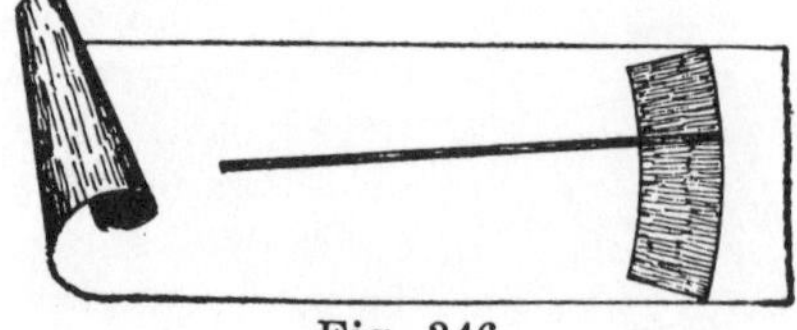

Fig. 346

radius pole in this case is on one side of the strip and not at the center of the strip of roofing, as in the other case. The narrow long shingles are suitable for rather steep roofs which do not need much lap. When the lap is narrow, it should be sealed with asphalt cement, and the end joints should be reinforced by slipping tin shingles under them, cementing the ends of the asphalt shingles to the tin shingles.

Stepping Off Odd Runs.—The problem that we are taking up here is most frequently met on lean-to roofs. That is, single-pitch roofs that are framed against some building or some other object. Whether or not the runs and the rises of such roofs are of even or odd feet and inches, does not matter, they can be stepped off accurately with this simple method:

Fig. 348 shows a lean-to rafter in place. The run in this case is $17\frac{3}{4}$ feet and the rise is $10\frac{5}{8}$ feet. With these figures to work with it would take some figuring to step off the rafter under the ordinary stepping-off rule.

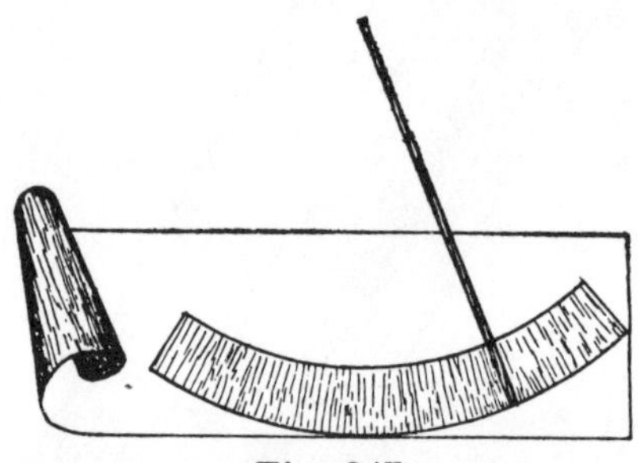

Fig. 347

But the stepping off in this case is even simpler than the ordinary stepping-off method under favorable circumstances.

If you will let the number of feet and fractions of feet be represented on the square by inches, which would mean $17\frac{3}{4}$ inches on the body of the square and $10\frac{5}{8}$ inches on the tongue, as shown by Fig. 349, and take twelve steps on the rafter material, you will have the exact length of the rafter. Putting it in the form of a rule:

Let the run and the rise in feet be represented, respectively, on the body and the tongue of the square in inches, and take twelve steps, which will give you the length of the rafter.

Divide and Conquer. — Fig. 350

shows a framing square applied to a straightedge with the figures 12 and 4½ intersecting with the edge. To the right we show a guide nailed to the straightedge, which should be thin material. The figures used on the blade and tongue of the square, as shown on the diagram, are always

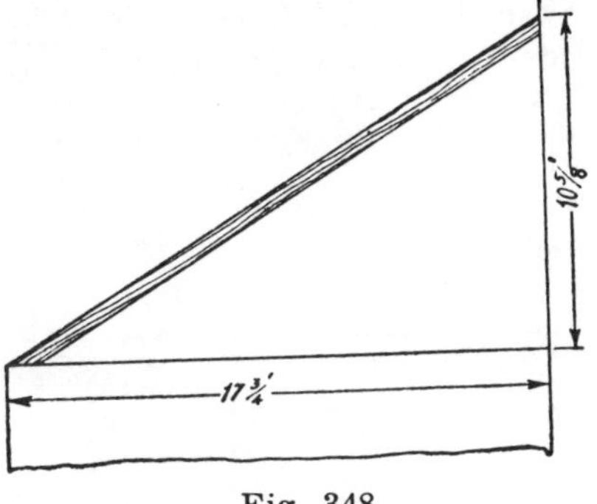

Fig. 348

read as inches, but the figures shown to the right between the arrows, are always read so as to make inches represent feet.

Now let us take the problem in division that we are starting with in this craft problem: Reading on the tongue, 4½ inches, between the arrows, 4½ feet and on the blade, 12 inches, which means that 4½ feet will have 12 4½-inch spaces in it. That is the problem

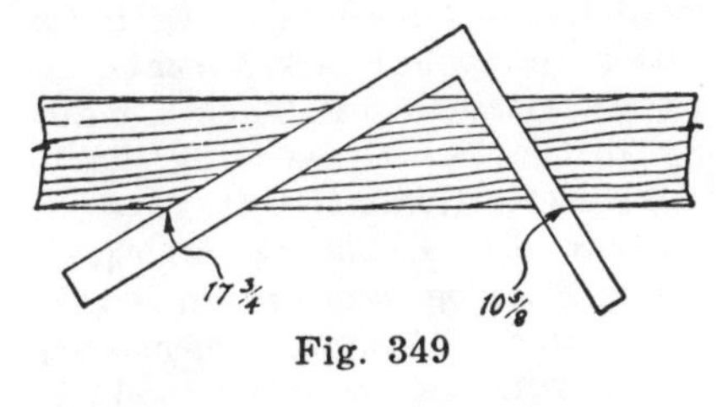

Fig. 349

we are using as a basis and would be stated this way: If 4½ feet have 12 4½-inch spaces, how many 4½-inch spaces are there in 8 feet? The problem is solved by moving the square from position *A* to position *B*, shown by dotted lines, and reading the intersecting figures, which in this case would be 21⅓, or 21 and ⅓ spaces. Another example: How many spaces of 4½ inches will there be in a distance of 9 feet? Move the square from position *A* to one inch past position *B*, indicated in part by dotted lines, and read the intersecting figure on the blade, which is 24, or 24 spaces.

We are purposely using figures that are easily divided so the student can prove the examples quickly. But in practice the figures that must be used in spacing will, in most cases be fractions. Let us say that we want to know how many spaces of 3$^{13}/_{16}$ inches there are in a distance of 5 feet, 3$^{7}/_{16}$ inches. We place the square

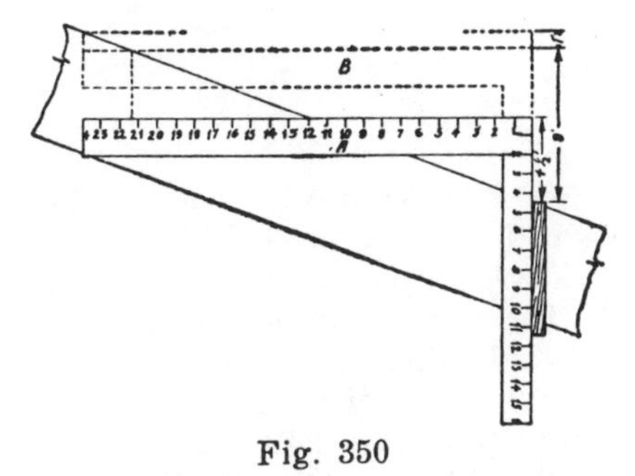

Fig. 350

on the straightedge, using 3$^{13}/_{16}$ on the tongue, and 12 on the blade, then move the square up to the point that represents 5 feet, 3$^{7}/_{16}$ inches, the intersecting figure on the blade gives us the exact number of spaces. (To simplify this, use the side of the square that has the inches divided into 12ths, or use decimals, and measure the fractions with a compass.)

One more example: If 4½ feet will have 12 4½-inch spaces, how many 4½-inch spaces will there be in 2¾ feet? The problem is solved by pulling the square back until the 2¾-inch point intersects with the edge of the straightedge. The intersecting figure on the blade gives the number of spaces.

The student should take a framing square and a straightedge and practice with them until he thoroughly understands the process of dividing distances with a framing square into spaces, using different figures for the length of the spaces.

Rapid Calculator.—Besides all of the other uses the steel square can be put to, it can be used as a board-foot calculator and the results are quick and accurate.

If the reader will turn to Fig. 351, he will find the steel square placed on a straightedge with a guide under it. The figures that are used are 12 and 4. In this position the square is set for calculating the number of board feet in 1x4's of different lengths. We are taking 1x4's because the board feet in such boards are easily figured, thus the reader can prove the accuracy of this rapid calculator. But the calculator will be equally accurate in giving the exact board feet for boards 5½ inches wide, 6⅞ inches wide or any other width, that might need to be calculated. The same is true with regard to the lengths of boards. They can be in even feet lengths, or they can be in lengths of feet and inches, including fractions, still this calculator will be accurate.

In the position the square is shown, Fig. 351, it gives the board feet for a board 12 feet long and 4 inches wide. The length of the board is shown on the blade of the square, where it intersects with the edge of the straightedge. To find the number of board feet in the board, read the figure where the tongue intersects with the edge of the straightedge; in this case it is 4, or 4 board feet. The original position of the square is marked *A*. Now, if the square is pushed forward to position B, shown by dotted lines, the intersection of the blade with the straightedge will read 15, and the intersection of the tongue will read 5, meaning that a board 15 feet long, 4 inches wide, will have 5 board feet of lumber in it. Or if the square were pulled back 3 inches, the intersections will read 9 and 3, meaning that a 1x4, 9 feet long, has 3 board feet of lumber in it.

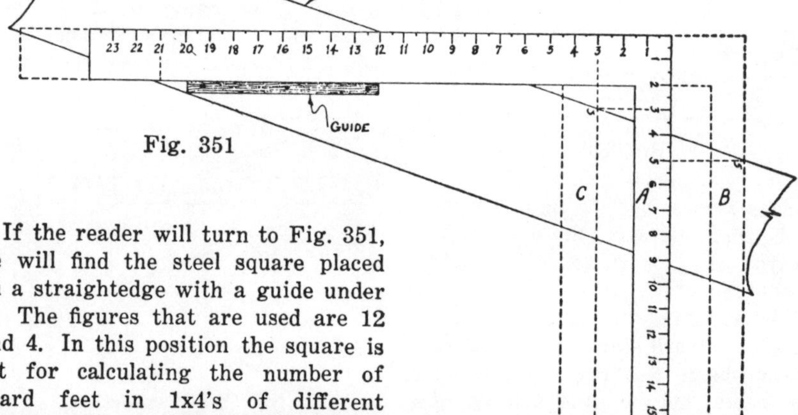

Fig. 351

The student should take his square and practice on even or odd lengths and widths of boards, remembering that the 12 on the blade and the width in inches, including fractions, on the tongue give the starting position. Then fasten the guide and shift the square according to the lengths of the boards, and the intersecting figures on the tongue will give the number of board feet of lumber in the various boards.

LESSON 35

PRACTICAL PROBLEMS

That Old Chestnut.—I happened to be in a group where a man presented the old chestnut of sawing an 8x8 block in such a way that it would

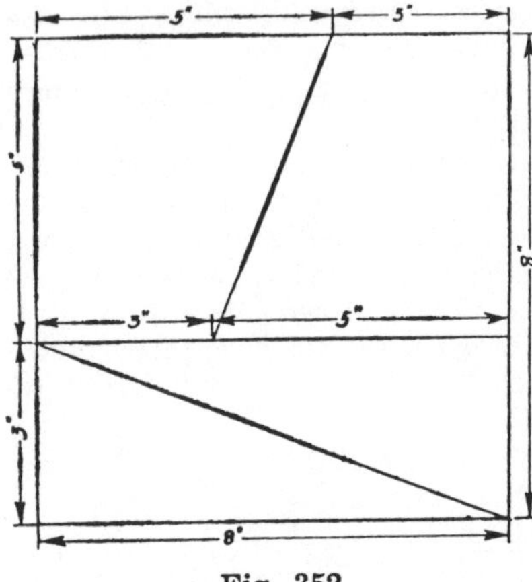

Fig. 352

gain one square inch in surface if placed in a certain position. He was not trying to fool anybody with it, because he actually believed that one square inch was gained; for 8 times 8 equals 64, or 64 square inches, while 13 times 5, the new position, equals 65, and therefore 65 square inches. But let me debunk the whole thing:

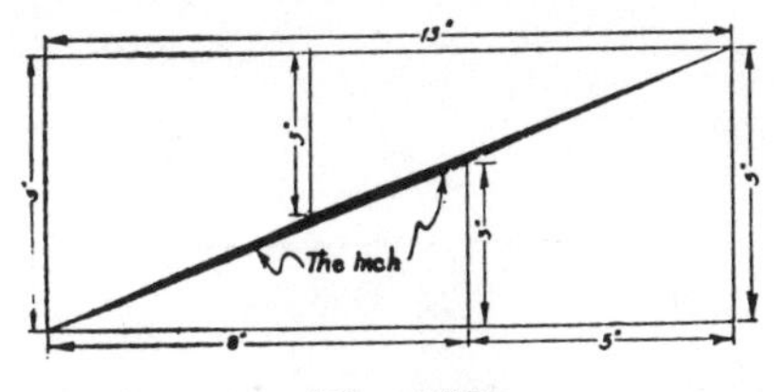

Fig. 353

Fig. 352 shows an 8x8 block marked for cutting into four pieces, which can be reassembled in such a manner that the new position, apparently, produces a surface having 65 square inches in it. Fig. 353 shows the 4 pieces reassembled into the 5x13 position, and anybody knows that 5 times 13 equals 65. . . . But here is the proof: If the reader will examine Fig. 353 closely, he will find that there is a heavily shaded diagonal streak that tapers off to a point at each end. That streak constitutes the alleged one square inch that was gained in sur-

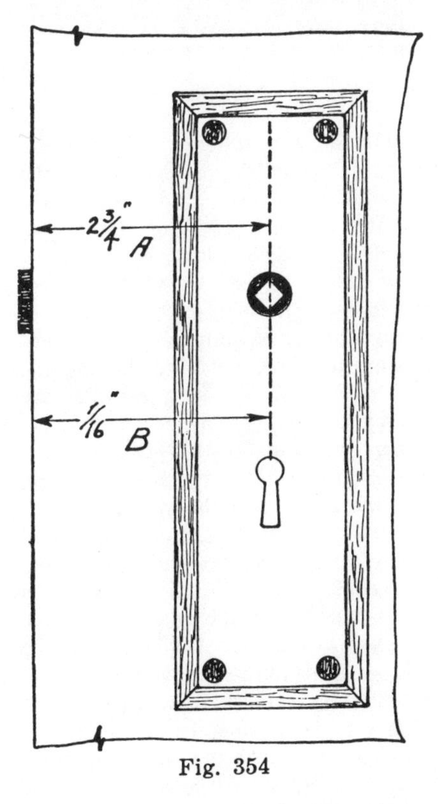

Fig. 354

face by the trick cutting and arrangement. But if you still are unconvinced, figure out the area of each of the four pieces and add up the number of square inches in them—the answer will be 64 square inches. . . . It's an old chestnut.

Lock Trouble.—A mortise lock that is properly installed should work freely—neither the latch bolt nor the spindle should bind (the binding is

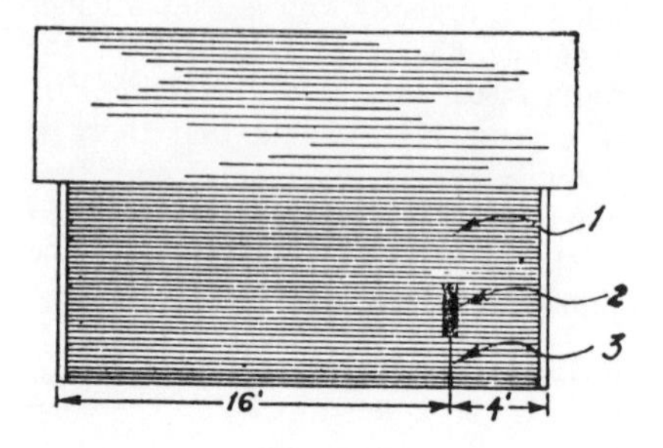

Fig. 355

usually on the spindle); if they do, there is something wrong with the installation. There is, however, a possibility of such binding developing after the work has been done. If the door stile is green or swollen when the lock is put in place, the stile can easily shrink enough to bring about a condition that will bind the spindle.

Fig. 354 shows how a stile can shrink as much as $\frac{1}{16}$ of an inch between the edge of the stile and the center-spindle line. When this occurs, the spindle is brought $\frac{1}{16}$ of an inch closer to the face of the lock, which could easily take up all of the play and bind the spindle. The remedy is easy—simply adjust the escutcheons in such a manner that the spindle will again have the proper relative position with the face of the lock. A stile that is thoroughly dry when the lock is installed and then becomes water soaked would develop the same trouble, excepting in reverse order.

Tin Over Joints.—Private garages often have to be extended in order to house automobiles that are longer than the ones for which the garage was originally built. Such an extension is shown by Fig. 355, to the right, where the joint is partially concealed with tin and paint.

Many of such garages are finished on the outside with novelty siding, and therefore we are showing novelty siding, although the tinning will work on any kind of siding. The arrow at 1 points to a part of the extension joint which has been tinned and painted, leaving the effect of a continuous run of siding. At 2, shaded, is shown a part of the joint tinned but still unpainted, while at 3 we have an untinned part of the joint.

Fig. 356 is a detail showing the joint in three different stages. At 1 we have two beads of a piece of novelty siding, showing how it will appear when the tin is on and painted with two coats of paint. At 2 we

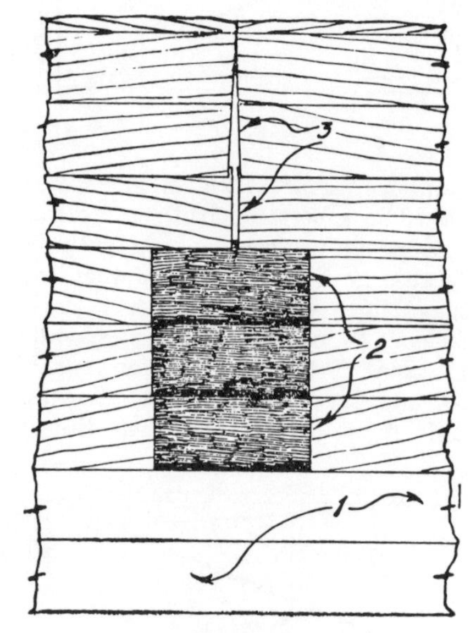

Fig. 356

are showing three beads of novelty siding, tinned, and at 3 we show the joint without tins or paint.

In order to secure a satisfactory job, the tins must be cut just the right size and bent to fit the mold of the siding. Then the tin is nailed to the siding with small nails or brads, not too many. After two coats of paint are on, the joint can not be observed from a little distance.

Fitting Sprung Moldings.—Most of the large moldings are sprung moldings, that is, such moldings are set on an angle with the wall, and there-

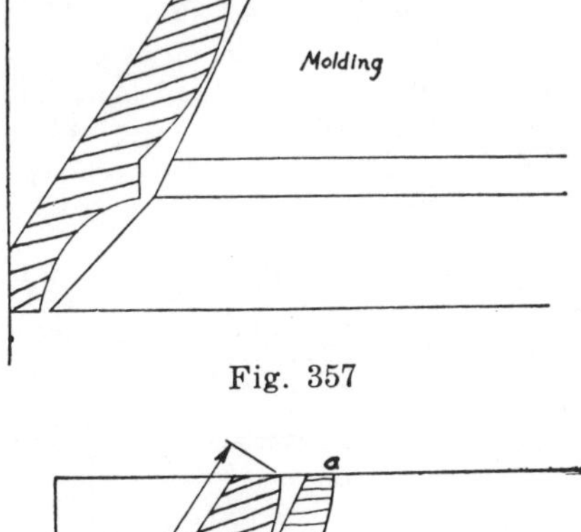

Fig. 357

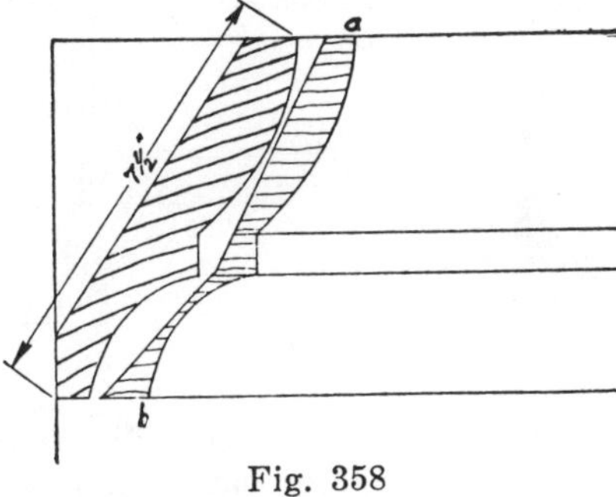

Fig. 358

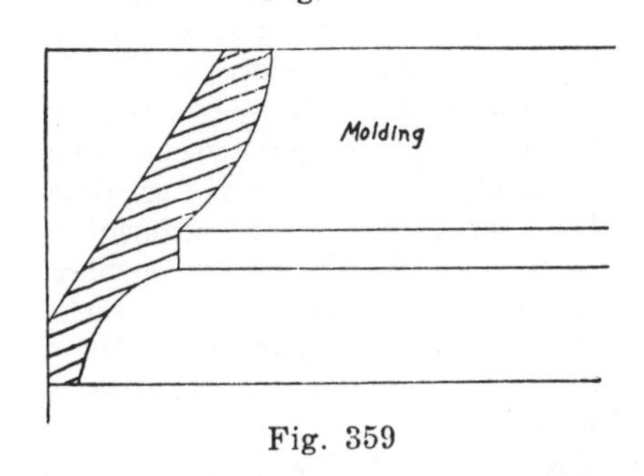

Fig. 359

fore leave an air space back of them in the angle. See the cross section shown in Fig. 357.

Many of the difficulties, that arise in coping sprung moldings, would disappear if the first piece of molding were left loose for several feet from the angle where the coped joint is to be made. Then when the coped piece is brought against it, it can usually be adjusted so that the joint will fit perfectly before the permanent nailing is done. When the first piece of molding is nailed before the coped joint is completed, the joint usually is open, either at the top or at the bottom. To make such a joint fit by cutting on the coped piece is

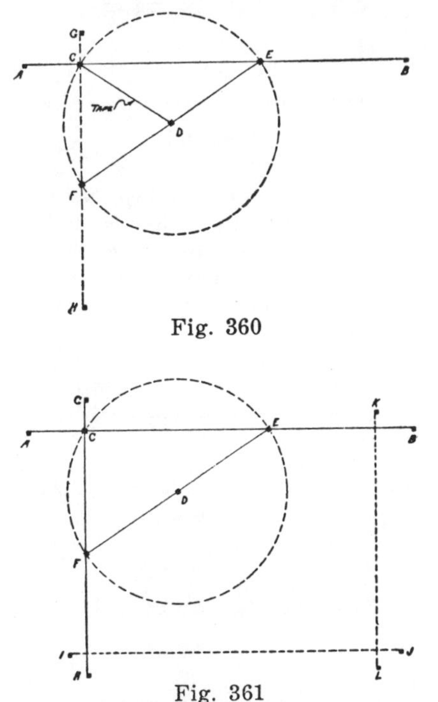

Fig. 360

Fig. 361

almost impossible, because the two pieces of molding will not member unless the first piece is set to the angle that will fit the coped piece.

In the case of extra large moldings, there is added another difficulty. They are too big to be cut in a standard miter box. Besides, such moldings sometimes are slightly warped, not enough to be noticed, but enough to cause trouble in making a coped joint. In such cases I usually tack the first molding in place, as shown to the left, Fig. 357. Sure that the first piece is in the right position, I give the end of the piece to be coped a rough cut, as shown to the right, and tack it in place about as shown. Then I take my scriber and scribe the mold-

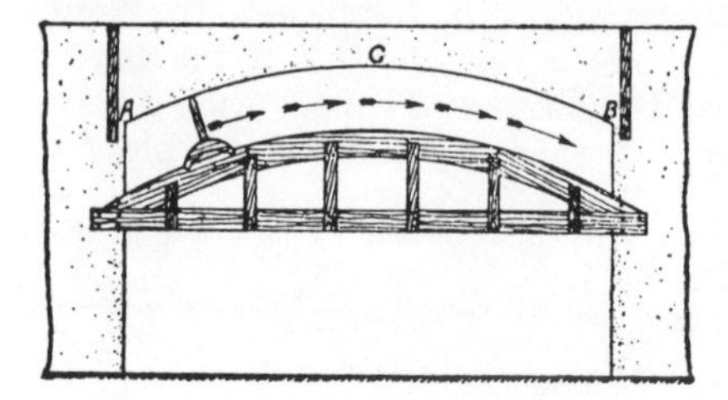

Fig. 362

ing as shown in Fig. 358, between *a* and *b*. The part that is lightly shaded is the part to be cut out. If this work is painstakingly done, the coped joint should fit as shown by Fig. 359.

A word of caution: Care must be taken that both pieces of molding are in the right position when the scribing is done.

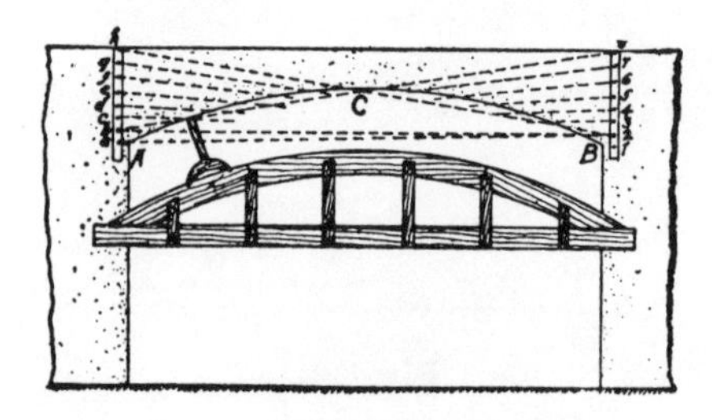

Fig. 363

Practical Squaring.—Fig. 360 shows the first operation of a most accurate and simple method of squaring a building. Stretch line *A-B* on the established building line. Then drive a stake, as at *C*, and establish the corner of the building on it, directly under the line. Now drive a stake at any convenient point, as at *D*, and drive a nail in the top of it. Hook your tape on that nail and take the point established at *C* on the tape, and swing around to *E*, driving a stake directly under where the circle crosses the building line, and stick a nail there. Fasten a line to this nail, and crossing *D* carry it past *F*. Now take up the tape again at *E* and swing around with it to *F*, as shown by the dotted circle—where the circle crosses line *E-F*, drive a stake and mark the point with a nail. With these points established, stretch line G-H, crossing *C* and *F*, and the angle at *C* will be a perfect right angle.

Now the dotted lines, *I-J* and *K-L*, Fig. 361, can be established by measurements; from these lines and the

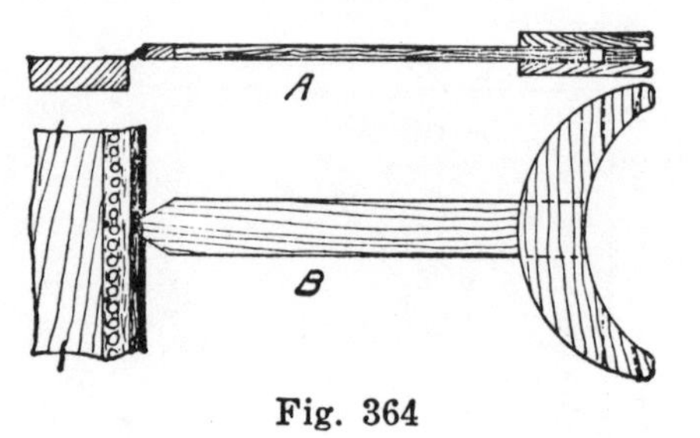

Fig. 364

two previously established, all other lines of the building can be located. There isn't a simpler or safer method of squaring small buildings.

Corner-Beading Segments.—It is an easy matter to put corner beads on small segments over openings where the radius of the segment is less than the height of the opening. But to put on corner beads over a wide segment-top opening without the use of a radius pole, is quite a different matter.

Fig. 362 shows a wide segment-top opening with a templet in place for gauging the corner bead with a pointer shown to the left. The arrows show how the pointer is moved to the right for gauging the corner bead while the nailing is being done. The nailing is started with the pointer at

A and carried on through *C* and to *B*. To keep the bead in line with the plaster surface, stretch a line from 2 to *a*, then from 2 to *b*, *c*, *d*, *e*, *f*, *g*, and *h*, as shown in Fig. 363. This will guide the bead to point *C*. Now stretch your line from *b* to 8, then from *b* to 7, 6, 5, 4, 3, 2, and 1. Of course, as you move from point to point with the line and pointer the nailing has to be done in order to bring the corner bead to a true segment, and keep it lining with the finish plaster line. To put on the other corner bead, transfer the segment templet to the other side of the opening and use the pointer to gauge the segment and with a gauge dapped to the width of the finished jamb, bring the bead to the plaster line.

Fig. 364 shows two views of the pointer. At *A* we have an edge view and at *B* a side view.

It should be remembered that the segment templet, which is a guide for the pointer, must be made with a radius the length of the pointer shorter than the radius for the segment to be formed. A little study of the drawings might be necessary to get the proper application of this method.

LESSON 36

TOOLS AND SAW FILING

Thimble Gauge.—Every carpenter has used what is known as the finger gauge for making short gauge marks; that is, the pencil is held with the index finger and thumb, while the other fingers are used for a guide, slipping along the edge of the material to be gauged. This method is all right on smooth material, when only short gauge marks are needed, but when long pieces of material are to be gauged, or material that has rough edges is used, then it is likely to either burn the finger tips or injure them with slivers. All of this can be avoided by using a thimble for the fingers, which makes it possible to gauge any lengths of material quickly and accurately without injury to the fingers.

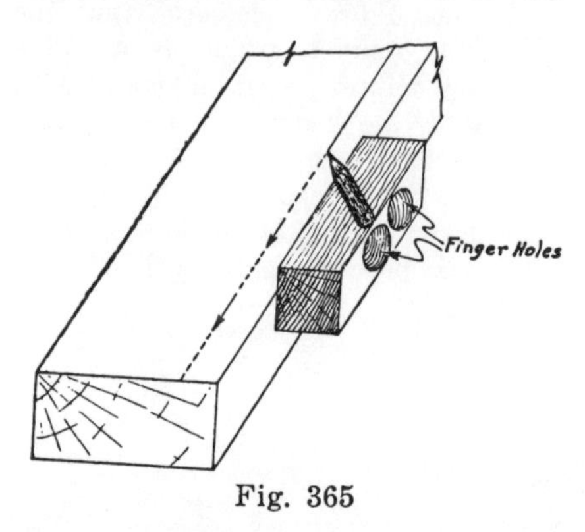

Fig. 365

Fig. 365 shows a perspective view of the thimble in the position for gauging, with the hand and fingers omitted. The two finger holes are just large enough so the fingers will fit into them snugly. Fig. 366 gives a view of the thimble, showing the position of the hand and how the pencil is held. A notch should be cut in the corner of the thimble, just right for the pencil to fit into.

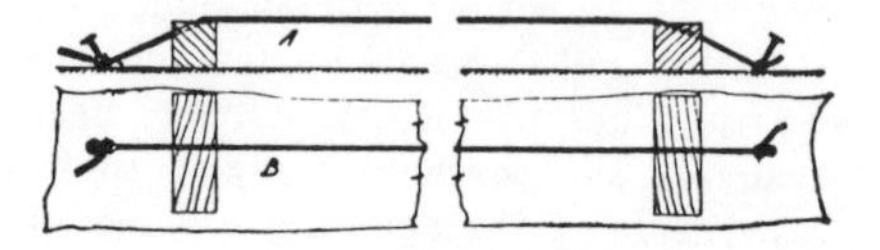

Fig. 367

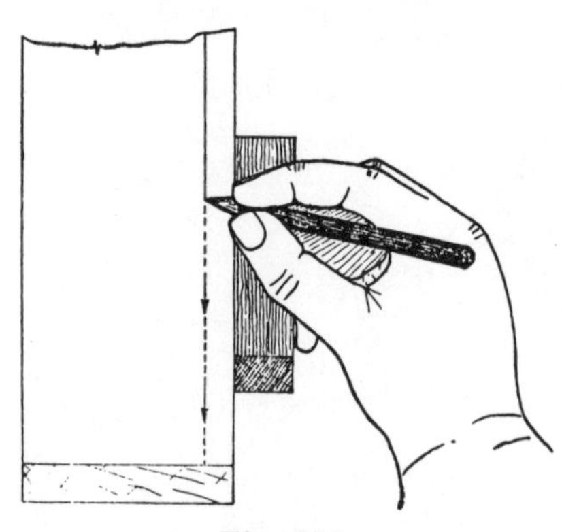

Fig. 366

The thimble is made, as the reader might have observed, of a block of wood 1x1 and about 3 inches long. The holes should be bored before the block is cut to size, in order to prevent splitting while the boring is done. When the holes have been bored, cut the block to the proper size and shape.

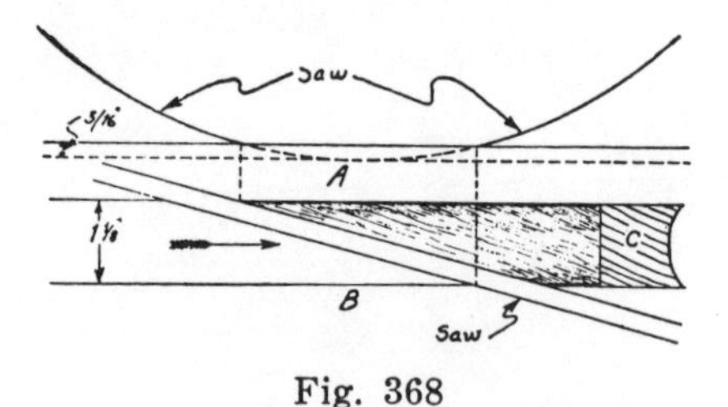

Fig. 368

Kerfing Offset Blocks.—Offsetting a line when it is used for straightening studding, forms, joists and so forth, is not a new thing. It is well known to carpenters everywhere, but kerfing the offset blocks to hold them in position is not so well known.

In offsetting a line, say, for straightening a side form for a concrete wall, the accepted procedure is to use ⅞-inch blocks and slip a block under the line at each end to hold the line away from the form far enough to miss the bulges or crooks. There are different ways employed to hold the blocks in place; some carpenters tack them to the form and others nail the blocks to the form with the nail left partly undriven to fasten the line to. These methods are all right, but the method shown by Fig. 367 is the simplest and most practical of them all.

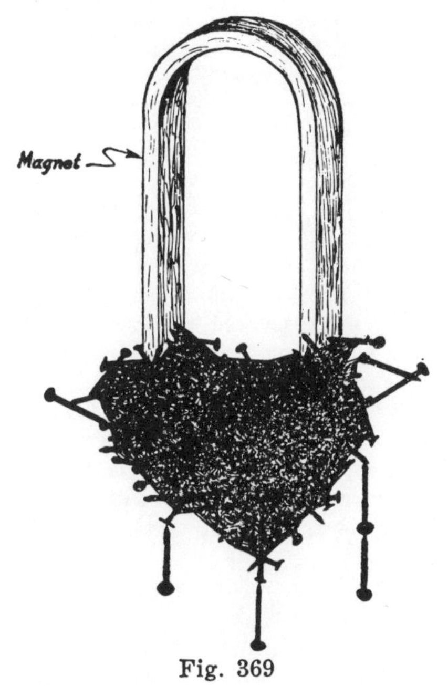

Fig. 369

At *A* is shown a line that is offset ⅞ of an inch, giving cross sections of the blocks, and showing by dotted lines how the blocks have been kerfed for the line to slip into. This holds the block in position. At *B* is shown the same line blocked out with the same blocks, but looking straight at them.

Making Grooves.—I was asked once whether a groove 3/16-inch deep and

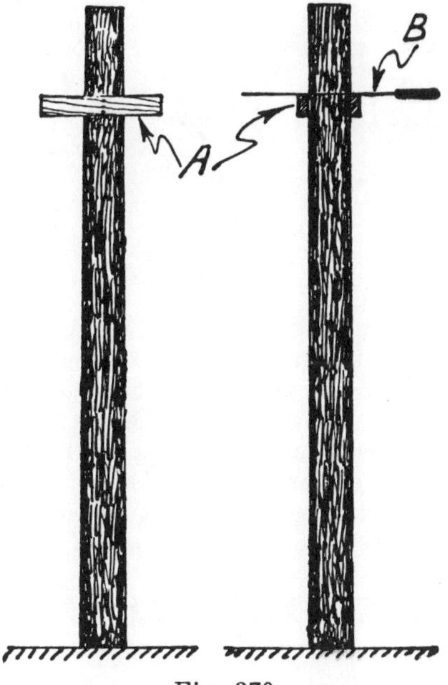

Fig. 370

1⅛ inches wide could be cut with a ¼-inch novelty table saw with one operation.

While I have worked around power-driven saws, and have used them some, I do not want to be regarded as expert in this field. However, Fig. 368 will prove that the groove can be cut with one operation. At *A* is shown by dotted line the depth of the groove, and I also show a segment of the saw. The dotted part of this segment gives a smaller segment of the saw—the segment that will cut the groove. At *B* is shown a plan of the elevation given at *A*. The two perpendicular dotted lines connect the two drawings and give the points for determining the angle of the saw for cutting the groove. The shaded part to the right shows the groove that is already cut. The arrow to the left shows the direction the material is pushed against the saw when the groove is made. At *C* is a cross section of the grooved piece.

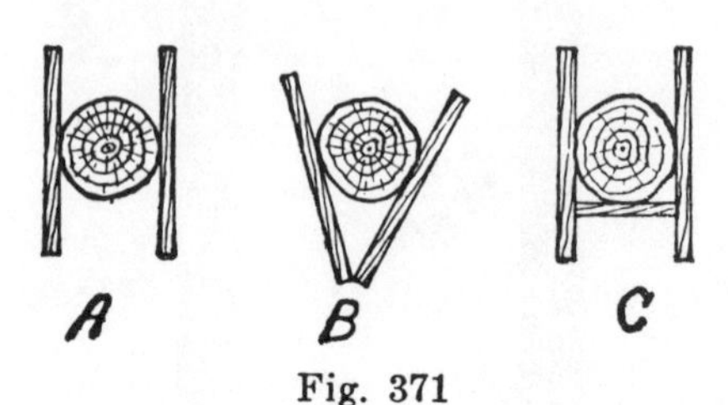

Fig. 371

A little study of the diagram will reveal the fact that if the depth of any groove is known, the angle to which the saw must be set to make the groove, can be determined by applying the principle used in the diagram for solving this problem. It should be remembered that in cases

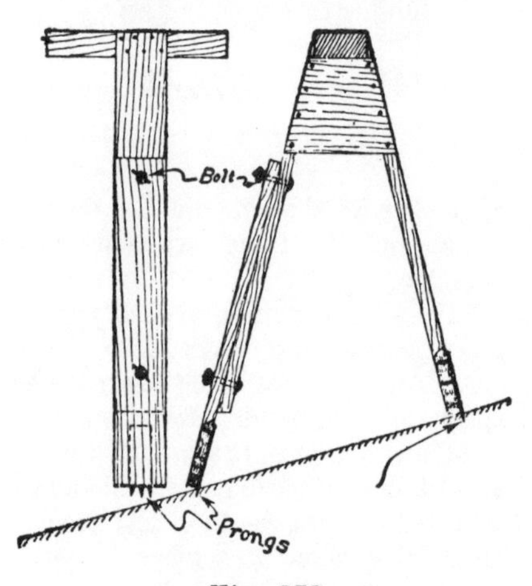

Fig. 372

where the groove is rather deep, that more than one operation is necessary.

Handy Tool.—One of the handiest tools a carpenter can carry in his handbox, especially if he is using small nails, screws, and so forth, is a fair sized magnet. Perhaps many of my readers already use the magnet in the way I shall suggest, but there are many who do not, and they are the ones who will benefit by this little trick.

Quite frequently one drops a nut or a screw or some other metal part of something he is working with that might be rather difficult to locate. If he has a magnet he can pass it over the space where the lost part is hidden, and in this way recover it without much waste of time. But

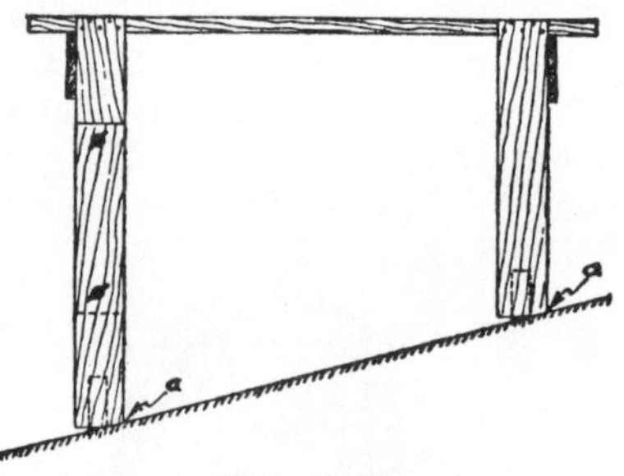

Fig. 373

should he be working with brads, tacks or even still smaller metal objects, and these are spilled, either on the floor or into dust, they can easily be recovered to the last one with a magnet, somewhat in the order shown by Fig. 369.

Cutting Round Posts on a Level.—It is a rather difficult job to cut off the tops of round posts or trees with a saw, so that the end will be perfectly level without some kind of guide.

Fig. 370 shows two views of a post which has two short 2x4's nailed to it at the point where it is to be sawed off. The first thing to do is to get the point where the top is to be cut off, and then nail one of the guides to this point in a perfectly level position. Then with the level transfer the point to the opposite side of the post and fasten the other guide there, also in a perfectly level position. These guides are pointed out at *A*. At *B* is shown the position of the handsaw and how the guides support it so the cut will be perfectly level when it is completed.

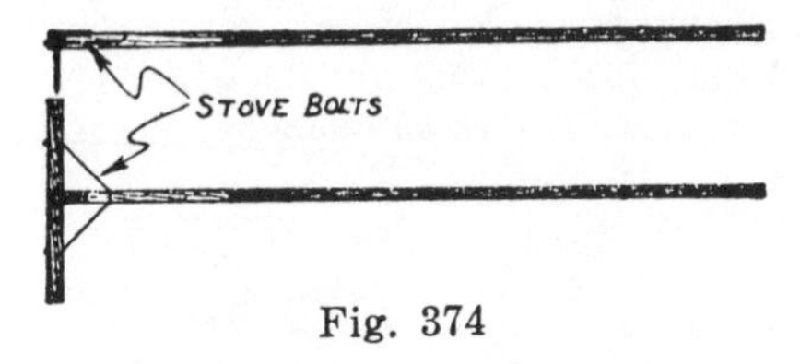

Fig. 374

Fig. 371 shows a plan of these guides at *A*, but this is not the only way that guides can be fastened. At *B* is shown a plan where the guides are fastened in a sort of V-shape, which has some advantages over what is shown at *A*. At *C* can be seen another method of fastening such guides. Here one of the long guides is fastened in its proper position and leveled. Then the short

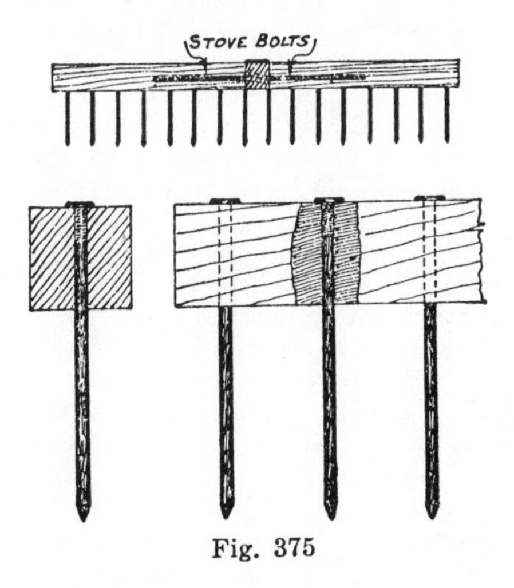

Fig. 375

piece that joins the two guides is leveled and nailed, leaving the other guide to be leveled and nailed. It should be kept in mind that the sides of the guides must be in a plumb position, when the guides are nailed, and the edges of the guides must be level both ways.

Tressel Crutches.—Working on inclines is often necessary and when a tressel has to be used it becomes a problem. In some instances the work has to be done, speaking of roofs, before the shingling is done. In other instances the problem must be solved after the shingles are on the roof.

How to put crutches on a tressel for use on inclines is shown by Figs. 372 and 373. The incline is rather flat, which was chosen intentionally to accommodate the drawings. In practice, both flat and steep inclines present themselves for solutions.

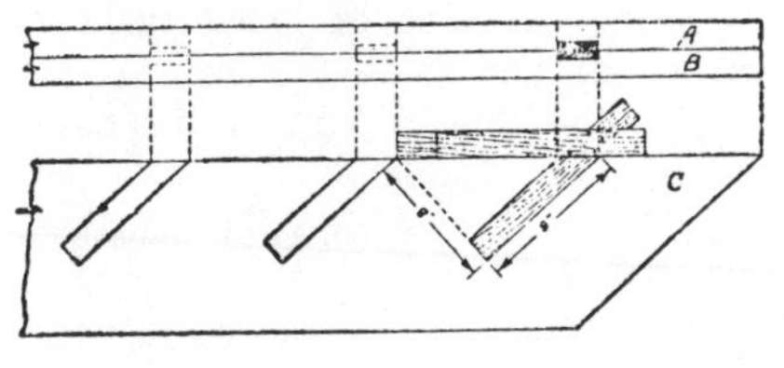

Fig. 376

Fig. 372 shows an end view of a tressel with one crutch on, and a side view of the tressel leg with the crutch. The crutch is shown bolted to the tressel leg with thumb bolts, which is quite satisfactory where the crutches are to be used at various times, or when they have to be changed frequently while in use. If, however, the crutches are to be used for one operation only, then they can be nailed to the legs, and when the job is finished they can be removed and thrown away.

The prongs pointed out at the bottom are somewhat exaggerated, especially for use on an incline as flat as the one shown. But if the incline is steep, then the prongs must be long enough to insure safety.

Fig. 373 shows a tressel, with crutches, at a right angle to what is shown in Fig. 372. Here the corners of the legs and the crutches, pointed out at *a, a,* must be cut off as shown.

Job-Made Rake.—For raking blocks, scraps, and rubbish of all kinds there is perhaps nothing better than the job-made rake shown in Fig. 374. At the top is a side view, showing the handle and the end of the piece that holds the teeth, while the bottom view is a plan of the same thing. The teeth are made of spikes, the size of which must be determined by the workman himself, for conditions must be kept in mind in determining the size of the spikes. The heavy dots in the toothed piece represent the heads of the spikes, and the braces, which are long stove bolts are pointed out with indicators. A piece of sheet iron is used to tie the toothed piece to the handle. The shaded part of the handle has been rounded.

Fig. 375 shows details of the toothed piece and how the nails are placed. The top detail shows the han-

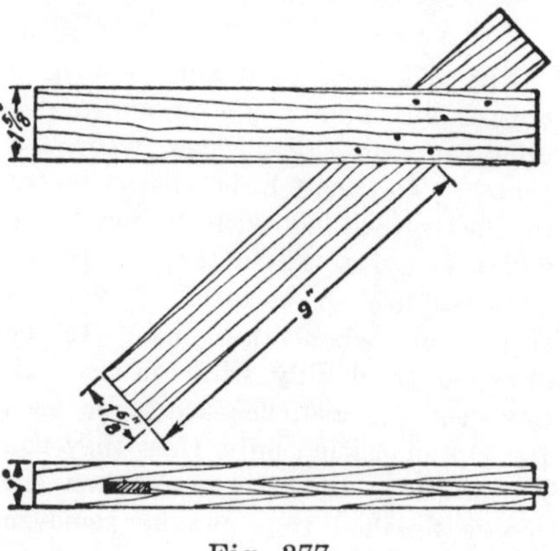

Fig. 377

dle cut off a little beyond where the stove bolts are fastened, which are again pointed out with indicators. The dotted lines on the cross section of the handle show how the holes for the stove bolts are slanted. The bottom details show, to the left, an end view of the toothed piece and the spike in position, while to the right a side view is shown. The center spike is shown with enough of the wood cut away to reveal clearly its position.

Housing Templet.—Plank stairs are usually made on the job and many of them are housed, which is to say that the treads are gained into the horses. *A* and *B*, Fig. 376, give the edge view of the two horses laid side by side for

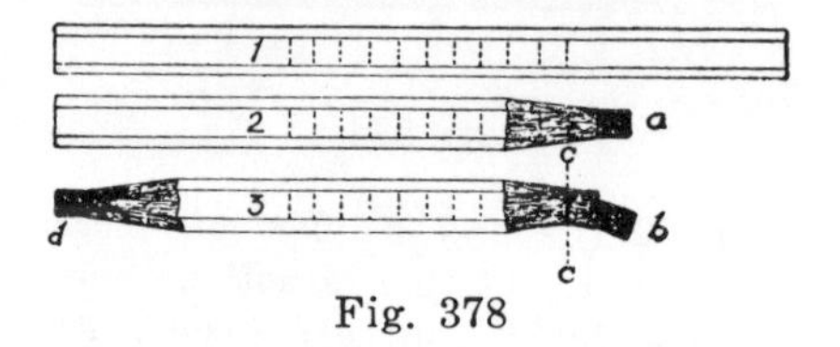

Fig. 378

marking the housing on the edges, while *C* is a side view of the horse marked *A*. The dotted lines running from *A-B* to *C* show the relationship of the two parts of the drawing. The housing to the right on *A-B* is shaded, which means that the housing is already done. The left and center housings are marked by dotted lines but not worked out. These two are shown marked on the side of the horse shown at *C*. A job-made templet is shown to the right applied to the horse ready for marking the housing. The templet is shown shaded and the dotted line to the left of the templet marks the place where it would have to be cut off in order to make it like the templet shown in Fig. 377, where the upper drawing gives a side view and the bottom drawing gives the edge view. The tongue of this templet, which is 9 inches long, is about ¼-inch thick and as wide as the thickness of the treads. To insure tight joints, the tongue should be two or three shavings narrower than the thickness of the treads. The length of the tongue should be governed by the width of the tread, which in this case is 9 inches. The tongue can be fastened to the guide with brads or with small screws, which are shown by the heavy dots. To mark the horse shown at *B*, Fig. 376, the other side of the templet will have to be used.

A Pencil Problem.—No doubt every carpenter has often found that when he sharpens his pencil, that toward the middle, the lead breaks. And when he sharpens it more the lead keeps on breaking. This keeps up until the pencil has been sharpened so much that it is too short for use. The truth of the matter is that the lead does not break while the sharpening is being done. It is already cracked, somewhat as shown by the dotted lines in Fig. 378.

Pencil number 1 shows by the dotted lines through the center how the lead is broken, either by the pencil being bent or for some other reason. When the pencil is used from the end, the lead holds until it is sharpened to about the point marked *a*, pencil number 2. At this point the pencil is

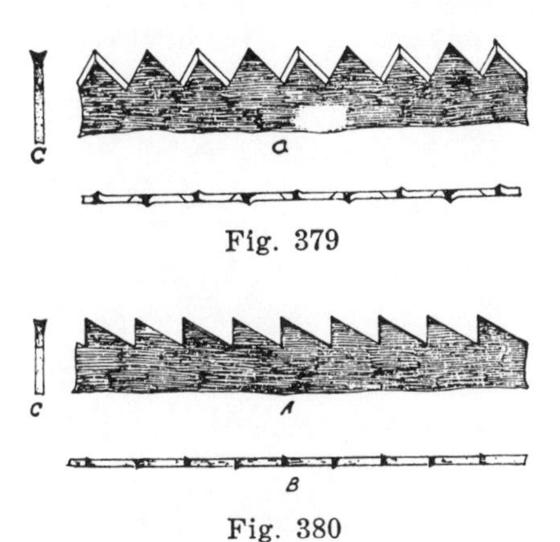

Fig. 379

Fig. 380

not too long and not too short, but then the lead begins to break somewhat as shown at *b*, number 3. When this happens, instead of resharpening the pencil, cut it off as shown by the dotted line between *c* and *c*, and sharpen the other end of the pencil, as at *d*. In this way you can utilize the best parts of the pencil without the lead breaking constantly, as it often does when resharpening is attempted through the center.

Saw Teeth and Sharpening Them.—The carpenter who can sharpen his saws so they will cut smoothly has found one of the principal keys to his

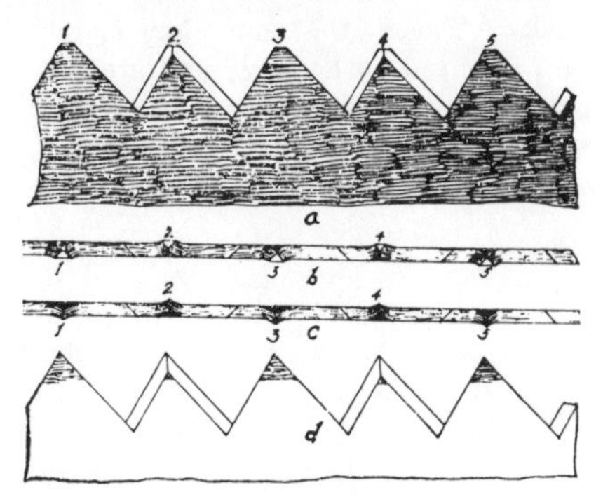

Fig. 381

success. This part of this lesson is to help those who might not have possession of that key now.

Fig. 379 shows at *a*, a side view of saw teeth of a cut-off saw. At the bottom are shown the same teeth, looking straight at the points. At *c* we have a section, showing the set of the points. In the same way, Fig. 380 shows the teeth of a rip saw respectively at *A*, *B*, and *C*. It will be noticed that a slight bevel on the teeth is shown, even though a rip saw should be filed straight across. However, in filing a rip saw, every other gullet should be filed from one side, then the saw should be turned and those that are left should be filed from the other side. This practice, in most cases, gives the teeth the slight bevel that is shown, which at the same time keeps the teeth balanced.

Fig. 381 shows four enlarged views of five saw teeth. At *a* is shown what happens to the points when a saw is jointed; at *b* is a view of the jointed points, looking straight at them, showing little white triangles at 1, 2, 3, 4, and 5. In filing a saw after it has been jointed, the saw should be so placed in the clamp that the light will strike the jointed points in such a manner that it will be reflected to the eyes of the saw-filer, in which case the points will appear as little white specks. These the saw-filer should keep in mind as he files, and stop filing just as soon as the white speck is gone. This, to get a first-class job, will often mean that the last stroke with the file will have to be only a part of a stroke. Another thing that should be watched is the shape of the teeth; for they should always be kept at a uniform size. If the filing develops, alternately, a large gullet and a small gullet, and so forth, then the saw-filer should regulate the pressure on the file so as to do the greater part of the filing on the large teeth. If this is watched, and painstakingly done, the saw teeth will look much on the order of those shown at *c* and *d*, Fig. 381.

LESSON 37

GROWTH OF WOOD AND SHRINKAGE

Trees Classified.—While carpentry does not directly include forests, trees, and woods, it nevertheless indirectly applies to all of them. And applying to trees, it must then also apply to the growth of trees. According to their growth, trees are put into two divisions, exogenous and endogenous. The exogenous trees grow by the formation of a new layer of wood each year outside of the existing wood, or between the existing wood and the

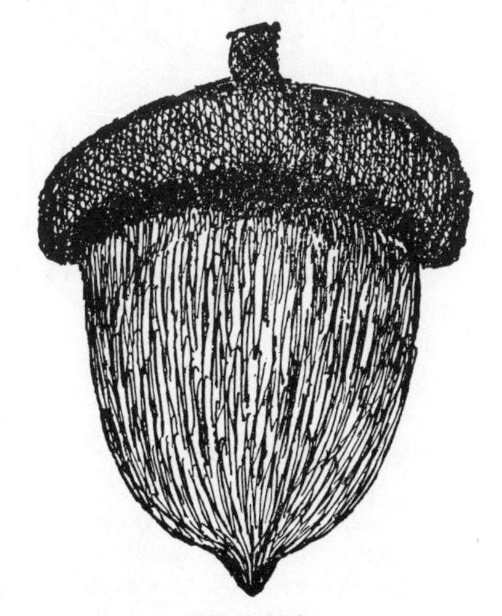

Fig. 382

bark. Ash, oak, pine, and walnut are examples of this kind of trees. The endogenous trees grow by the formation of the woody matter inside of the stem. Bamboo, Palmetto, and cane belong to this class. The exogenous trees, however, are the trees that produce the greater part of the material used in carpentry. These trees are divided into two classes, the broad-leaved trees and the needle-leaved. Oak, walnut, basswood, and ash are examples of the former class, while pine and cedar belong to the latter. On the other hand, the endogenous trees are also divided into two

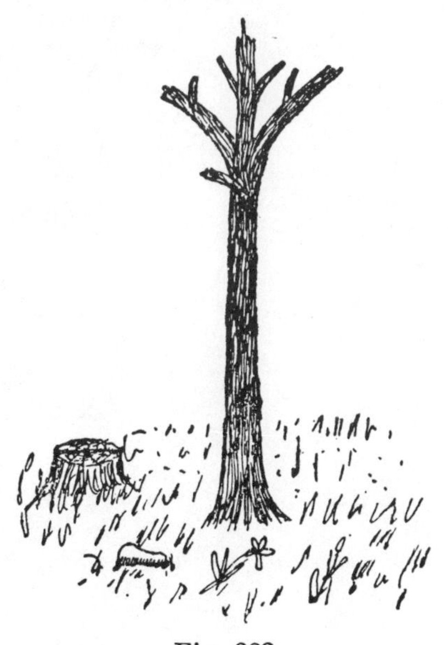

Fig. 383

classes; the bamboo being an example of the one class, and the palmetto of the other.

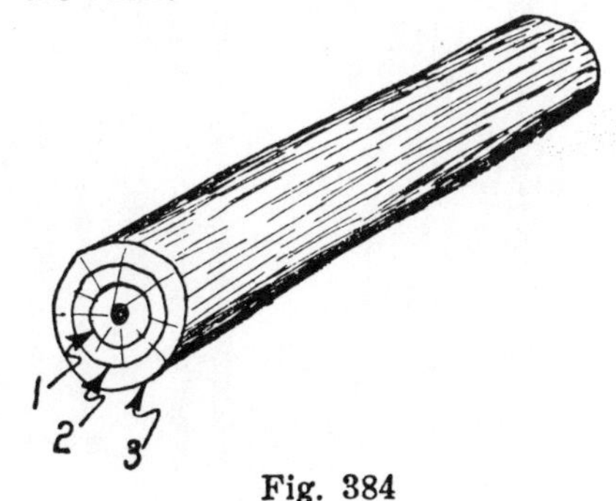

Fig. 384

The Giant Oak.—Among the trees, the oak is perhaps the most widely distributed tree, has the widest field of usefulness, and has the greatest number of different kinds or species, running all the way from a sort of shrub to the giant white oak. A large size acorn, the seed from which the

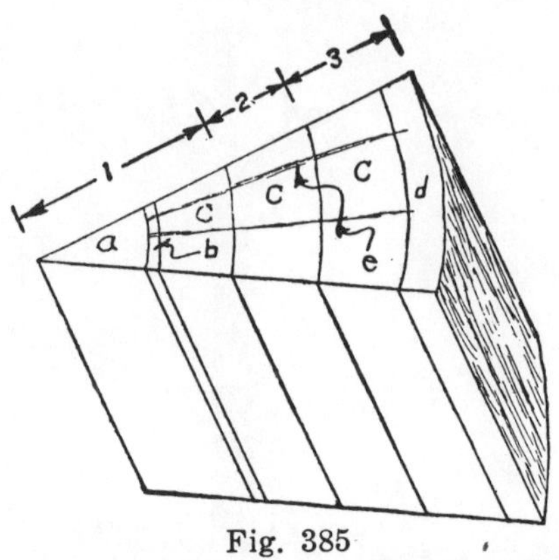

Fig. 385

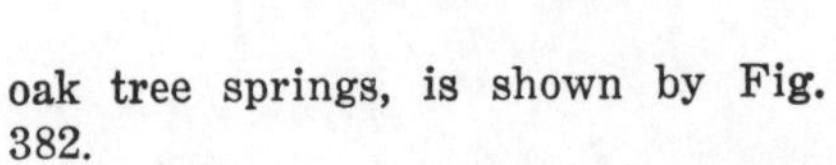

oak tree springs, is shown by Fig. 382.

Composition of Wood.—Fig. 383 shows a tree about three years old.

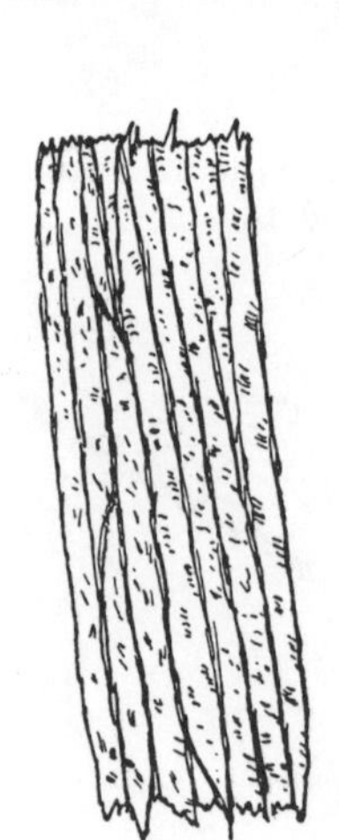

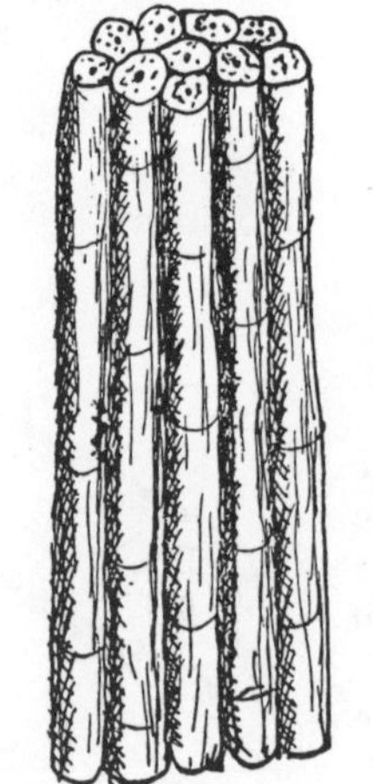

Fig. 386

To the left is shown a stump of another tree that has been cut down. A part of the stem of this tree is shown by Fig. 384. The shaded circle at the center represents the heart. The first year's growth is shown within the circle pointed out at 1, the second year's growth is shown between 1 and 2, and the third year's growth, between 2 and 3.

Fig. 385 shows a wedge-shaped enlarged part of the stem of a three-

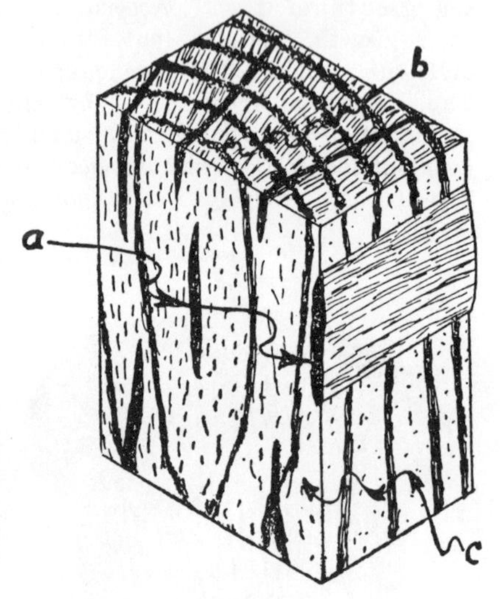

Fig. 387

year-old tree. The three years' growth is indicated at 1, 2, and 3. The bark is shown within the third year's growth, although it has been produced during the three years of the tree's life. If the student would take a magnifying glass

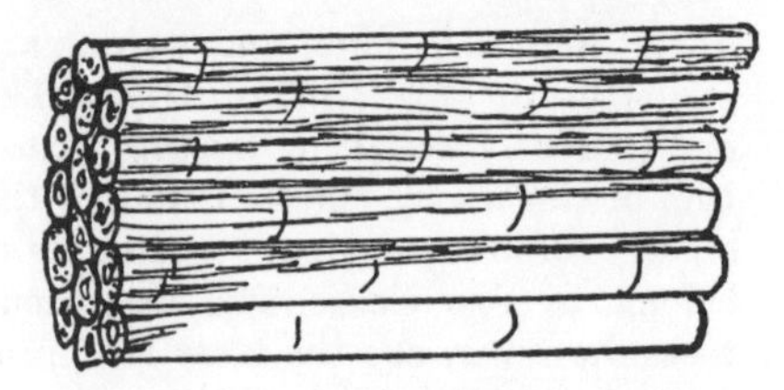

Fig. 388

and hold it over a piece of oak cut somewhat on the order of what is shown by Fig. 385, he would find many interesting things. At the part marked *a* he would find a netlike fabric of cells, called medulla, or pith. At *b* he would find the medullary sheath which surrounds the pith. This constitutes the inner layer of the first year's growth, and is composed of spiral vessels and fiber ducts for the purpose of carrying the sap to the

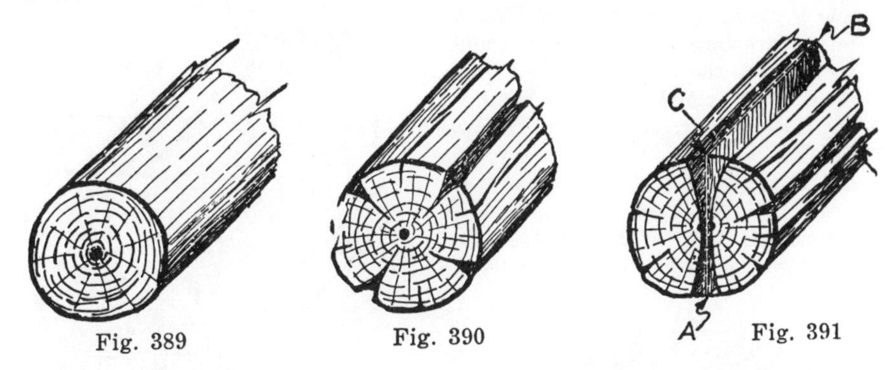

Fig. 389 Fig. 390 Fig. 391

twigs, where nature uses it in building up the tree. At *c, c, c* the student would see the three yearly growths of wood fiber, or fiber tubes, which are utilized by nature in conveying sap. The inner part of each

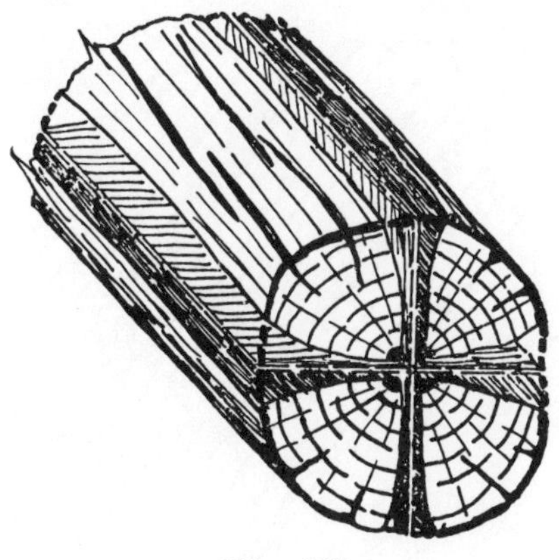

Fig. 392

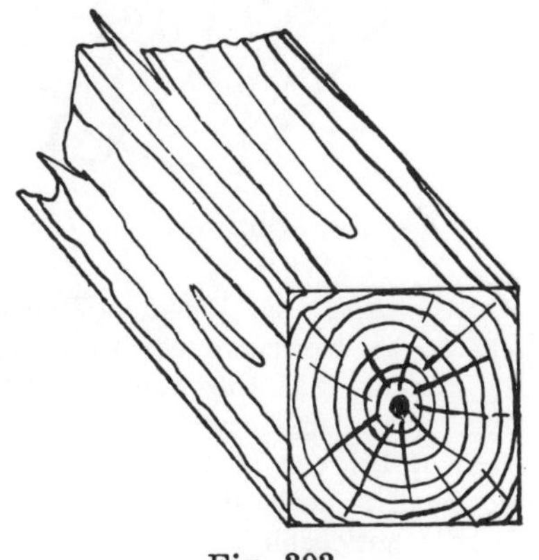

Fig. 393

yearly growth, composed of coarse fiber tubes, is called the spring growth, and the outer part, composed of much finer tubes, is called the summer growth. At *d* he would find the bark, composed of three layers. The inner layer is composed of woody fiber combined with a cellular tissue. The middle layer is composed of prismatic cells and tubes filled with juices, while the outer layer is of a dry corky composition. At *e* the student would see what is called the medullary rays.

Fig. 386, to the left, shows an enlarged drawing of what is called woody fiber, or vascular tissue, which is in the form of long tapering tubes, interlaced and breaking joints with each other, having a small portion of cellular tissue at intervals. An enlarged drawing of cellular tissue is shown to the right. These drawings are given merely to give the reader a better idea of what is meant by the terms, "woody fiber," and "cellular tissue." Examining a piece of oak through a magnifying glass, of course, will give the student the clearest idea.

Fig. 387, somewhat enlarged, shows, *a*, the medullary rays; *b*, the spring growth, or spring wood; *c*, the summer growth, or summer wood. An enlarged drawing of a bundle of medullary rays is shown by Fig. 388. When these rays extend from the heart to the bark they are called primary rays, but where they extend only through a portion of the distance between the heart and the bark, they are called secondary rays. Medullary rays are often called, silver grain.

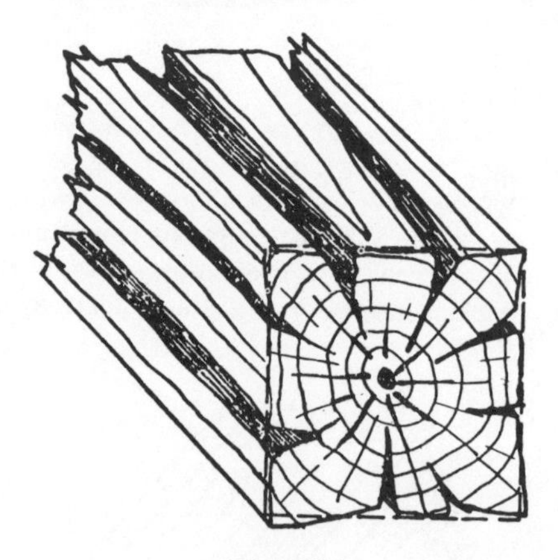

Fig. 394

Shrinkage in Wood.—Whenever checks or cracks appear in wood, it is evident that the wood is shrinking. Shrinkage in wood, in the direction of the run of the wood fiber, is so small in most instances, that it is hardly noticeable. The same thing is true relative to the medullary rays, especially in wood from broad-leaved trees. The wood from needle-leaved trees shrink more evenly, and consequently warp less than that of the broad-leaved varieties.

Fig. 389 represents a log that is still in its green state. The medullary rays are shown here radiating from the heart of the log, much like the spokes of a wheel radiate from the hub. Since these rays, speaking with reference to hard wood, shrink but little from end to end, it

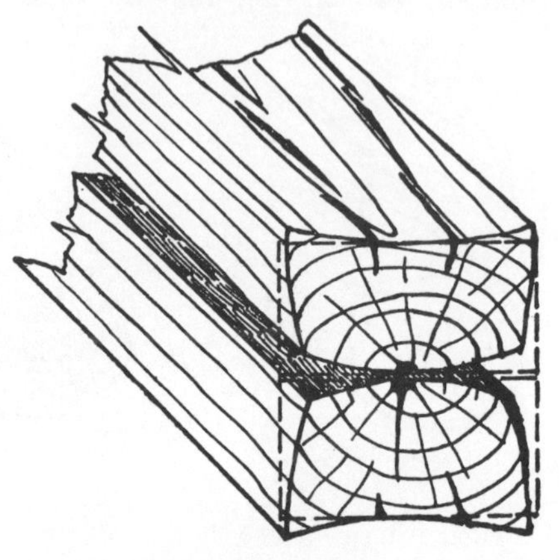

Fig. 395

will readily be understood why there is practically no shrinkage between the heart of a log and the bark. But

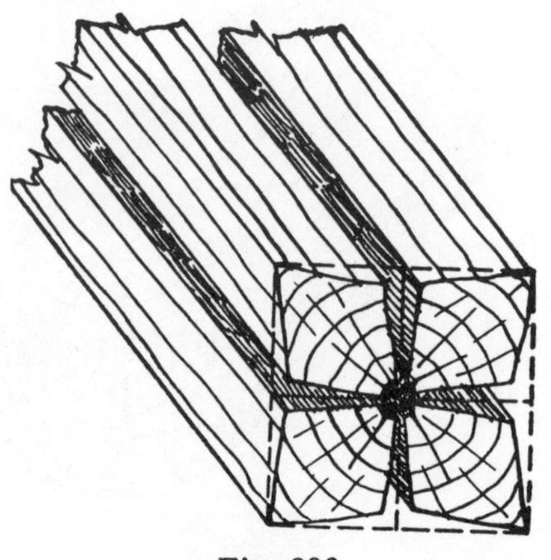

Fig. 396

when those innumerable little tubes of wood fiber become smaller, as the seasoning process continues, something is bound to happen, and that happening shows itself in the form of checks and later cracks. The end of a well-seasoned log is shown by Fig. 390; the cracks, however, have been exaggerated somewhat, in order to make the point clear. There are four rather large cracks shown and some minor ones. If the reader will imagine the medullary rays, shown on this figure, as being stays in a fan, it will not be hard for him to see that shrinkage in a seasoning log is much on the order of closing a fan slowly. It should be clear now, that the greatest amount of shrinkage in wood, viewing it from the end, is in the direction of the annual rings, or around the circumference of the log.

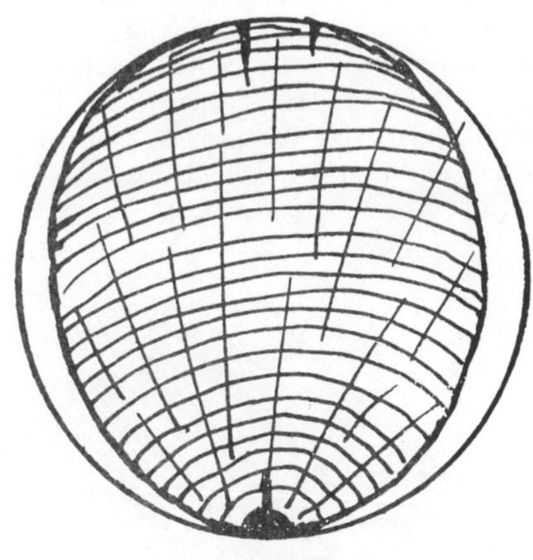

Fig. 397

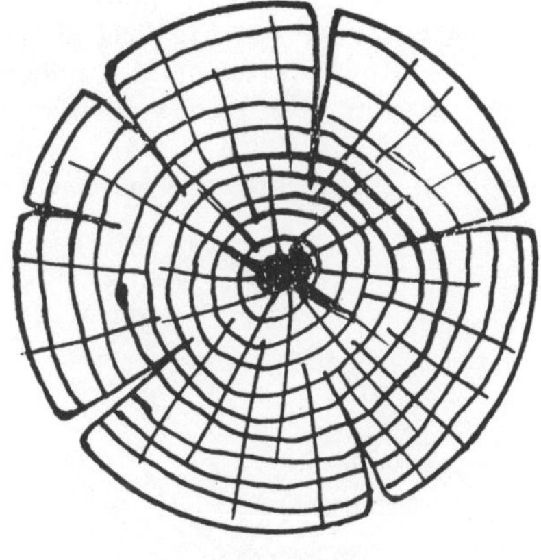

Fig. 399

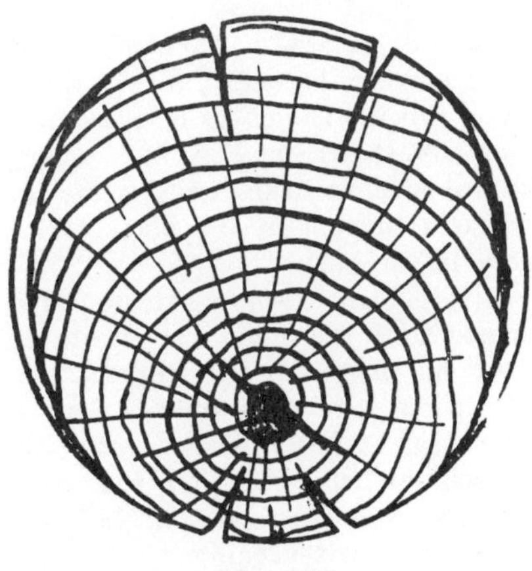

Fig. 398

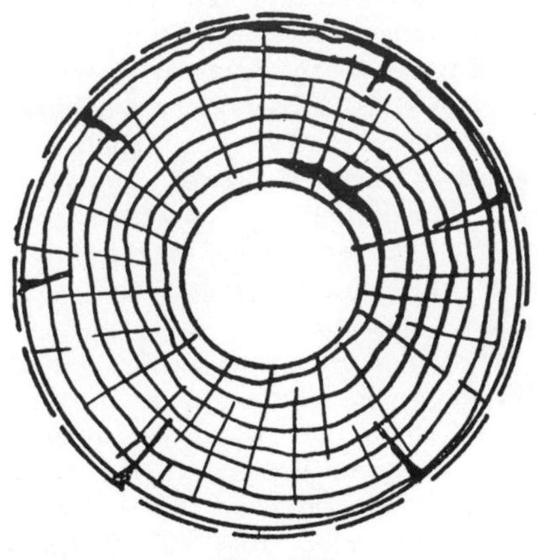

Fig. 400

Fig. 391 shows a log which was sawed into halves, while in its green state, as pointed out by indicators at *A*, *B*, and *C*. The fan-closing shrinking process is clearly shown here. The dotted lines show the shape of the two halves before they were affected by shrinkage. Notice that the cracks shown are somewhat smaller than those shown in Fig. 390. The end of a log sawed into quarters is shown by Fig. 392. The dotted lines show the shapes of the quarters immediately after they were sawed. The cracks here are still smaller than in the previous figure.

Fig. 393 shows the end of a piece of timber sawed into a square while the wood was green. Fig. 394 shows the same timber after it had undergone seasoning. Notice how shrinkage has warped its shape. (The cracks have been exaggerated for emphasis.)

Fig. 395 shows the timber, shown by Fig. 393, sawed into halves and after it had undergone seasoning. The dotted lines show the original shapes of the two ends. Observe the smallness of the cracks. The same timber is shown in Fig. 396, sawed into quarters. The dotted lines show the shapes of the quarters while in the green state. Seasoning has left the wood without checks.

Shrinkage on Round Timber.—Fig. 397 shows the effects of seasoning on a solid round post, which was shaped, while the wood was green, in such a manner that the heart of the wood was to one side. Fig. 398 shows a round solid post with the heart nearer the center. This post is more nearly round, but shows larger cracks than the one shown by Fig. 397. A round post with the heart in the center is shown by Fig. 399. Seasoning did not affect the diameter, but it produced large cracks in the wood. Fig. 400 shows a round post with the heart bored out. This provides ventilation for the center of the post, and it permits uniform seasoning. The results are that the post is perfectly round and almost without cracks. The dotted circle indicates that the post is somewhat smaller, due to shrinkage.

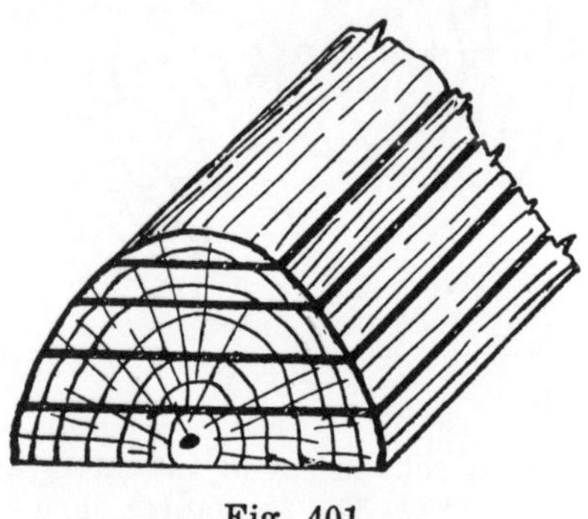

Fig. 401

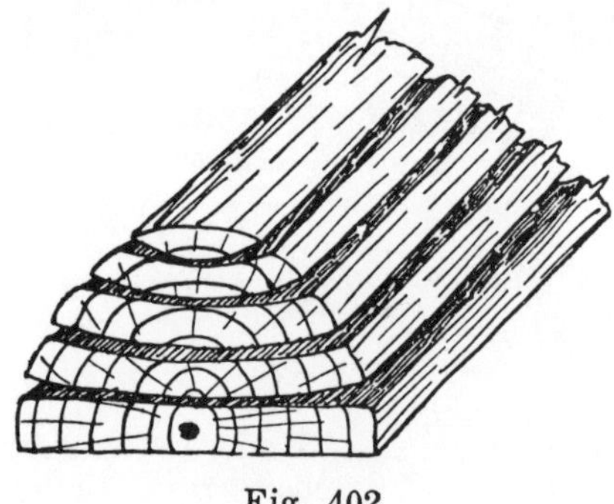

Fig. 402

Fig. 401 shows a part of a log sawed into five log-width planks while it was green. The effect of seasoning is clearly shown by Fig. 402, where the warping, from the heart outward, increases the nearer the plank is located to the bark.

It is evident from what we have learned thus far about the effects of seasoning on wood, that the manner of sawing logs into lumber has a great deal to do with the usefulness of such lumber. For that reason it is to the carpenter's benefit to know the effects of seasoning on lumber. For then he will be able to install it so that the effects of shrinkage will be kept at a minimum.

LESSON 38

SAWING LUMBER AND DEFECTS IN LUMBER

Slicing Logs.—So that lumber will give its greatest value in service, it is necessary to saw it in such a manner that it will answer the purpose for which it is to be used. There are different methods of sawing lumber. Rough lumber is usually sawed by the slicing method. This method is shown by Figs. 401 and 402. The same method is illustrated by Fig. 403. The slicing method of sawing lumber is also known as bastard sawing. After the log has been sliced, the slices are then sawed into whatever kind of lumber that the management wants. To the left of the drawing are shown, shaded, three planks sawed from two slices. Such lumber, as 2x4's, 2x8's, 2x6's, and other dimensions are usually sawed by the slicing method. This is the cheapest method of sawing lumber.

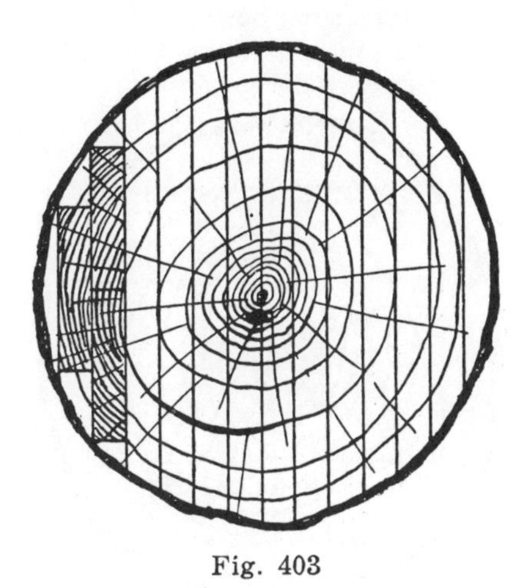

Fig. 403

Quarter Sawing.—Fig. 404 shows how a log is sawed so that the lumber will be what is called quarter-sawed lumber. It is also called rift-sawed, vertical grain, and edge grain. First the log is sawed into quarters, and then the quarters are sawed at angles of 45 degrees with the sawed sides of the quarters. This can be seen by the drawing. Then the wide pieces are trimmed and sawed into the desired lumber. The shaded pieces show

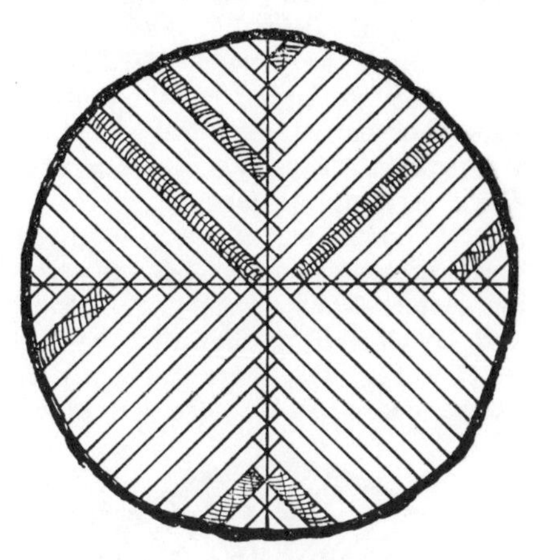

Fig. 404

the relative direction of the annual rings as they will appear on the ends of quarter-sawed lumber.

Four methods of sawing lumber are shown by Fig. 405. In each of these methods the log is first sawed into quarters. At *A* is shown the quarter-sawing method, which was explained under Fig. 404. At *B* is shown a method that brings out the grain a little more evenly than simple quarter-sawing does. A method of sawing lumber that gives good results is shown at *C*. This method is more wasteful than the two just explained. At *D* is another wasteful method, but the lumber sawed by this method is more nearly vertical-grained than the lumber sawed by the three other methods. The shading in this drawing gives the run of the annual rings on the ends of the different pieces.

Other Ways of Sawing Lumber.—Fig. 406 shows an economical way of sawing lumber, which gives good results, especially for sawing flooring. The direction of the annual rings can be determined by the shaded pieces. Fig. 407 shows a method of sawing large timbers. The dotted lines show further possibilities.

Marking for Heavy Timbers.—Three ways of laying out the ends of logs to be sawed into heavy timbers are shown by Figs. 408, 409 and 410. After striking as large a circle as possible on the end of a log, with the heart as center, draw *A-B*, crossing the center. Then at a right angle to *A-B*, lay off *F-E*, as shown by the dotted line, Fig. 408. This makes a perfectly square timber. To obtain an

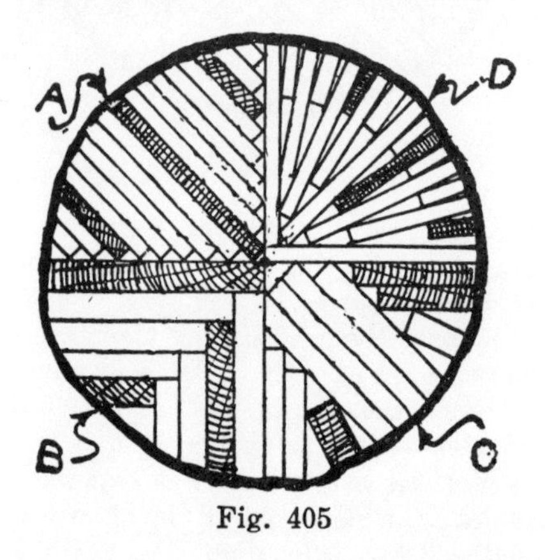

Fig. 405

oblong cross section of a timber, divide *A-B* into any number of equal parts—Fig. 409 shows three equal parts and Fig. 410 shows four. At a right angle with *A-B*, strike *D-E* and *C-F*. Now join, in each of the three illustrations, *A* with *F*, and *F* with *B*, and *B* with *E*, and *E* with *A*, and you will have the outlines of the ends of the three timbers. By studying these diagrams, it will be found that by increasing the number of divisions in line A-B, the outline becomes narrower and longer, but the heart of the log will always be approxi-

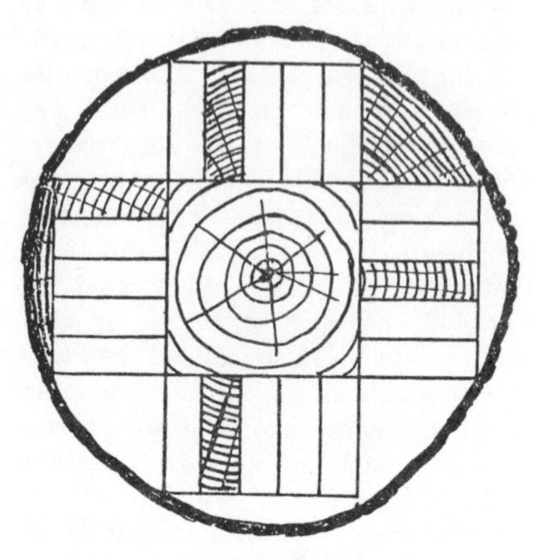
Fig. 406

mately at the center of the cross section.

Joining Timbers and Annual Rings.—Fig. 411 shows how boards should be joined in regard to the annual

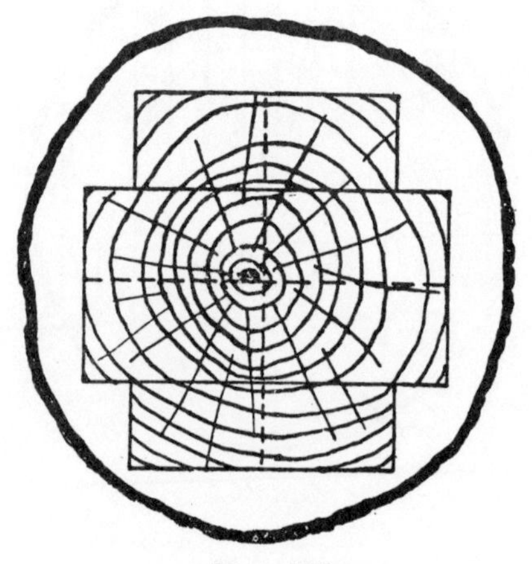
Fig. 407

rings, in order to counteract warping and shrinking. This method gives good results when the pieces are to be glued together. Fig. 412 shows how large timbers should be

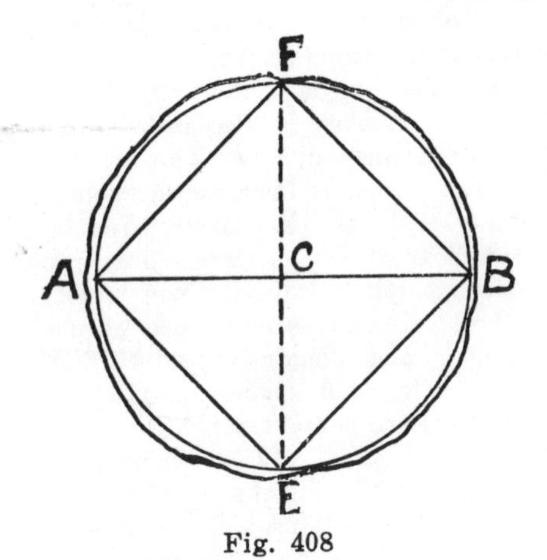

Fig. 408

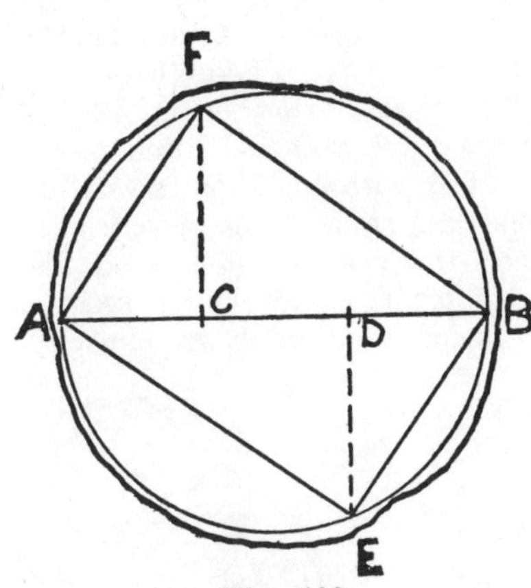

Fig. 409

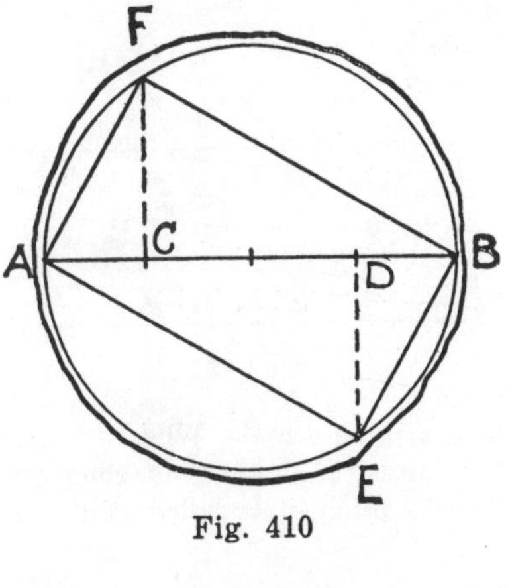

Fig. 410

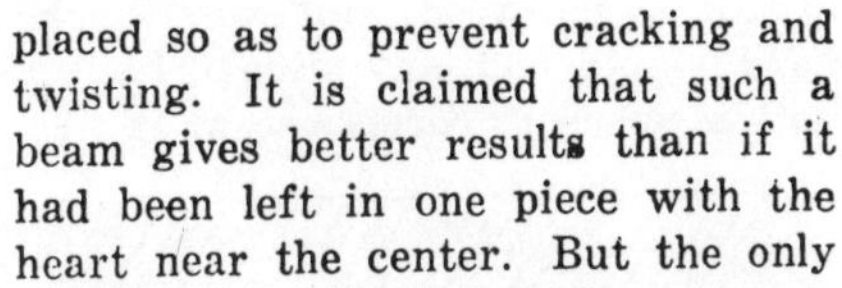
placed so as to prevent cracking and twisting. It is claimed that such a beam gives better results than if it had been left in one piece with the heart near the center. But the only

Fig. 411

thing that I can see about this beam that is better than if it were left in one piece, is that sawing it in two will reveal any defects that might be

Fig. 412

at the heart of the timber. This is of enough importance to justify sawing the timber in two.

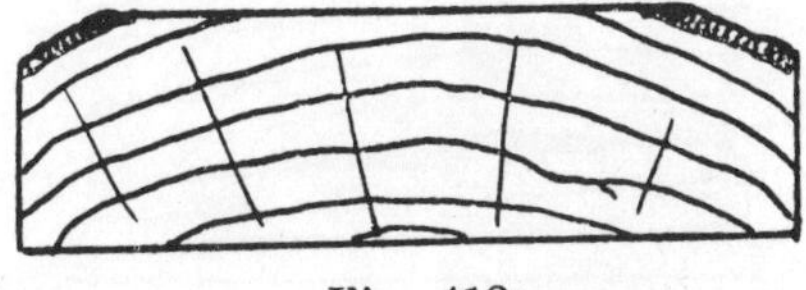
Fig. 413

Waney-Edged Lumber.—Fig. 413 shows a piece of waney-edged lumber. The two upper corners of the piece are not full size, and the bark is still on the wood. But whether the bark is on or not, any piece of lumber with a corner or corners scant in size, somewhat on the order shown, is waney-edged lumber.

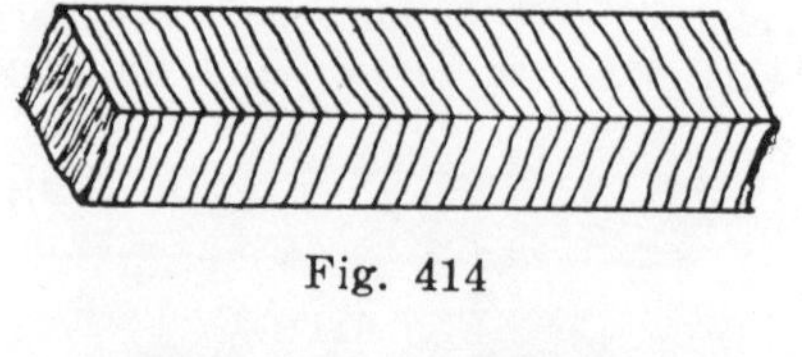

Fig. 414

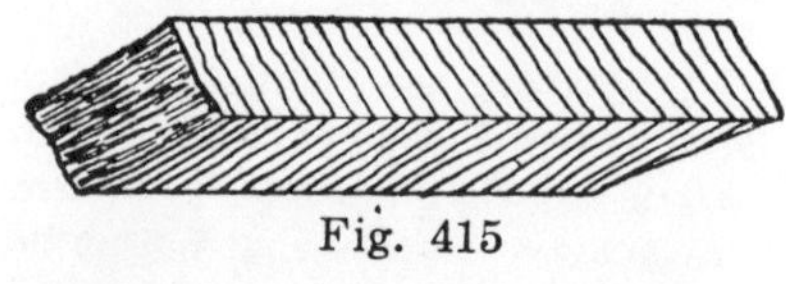

Fig. 415

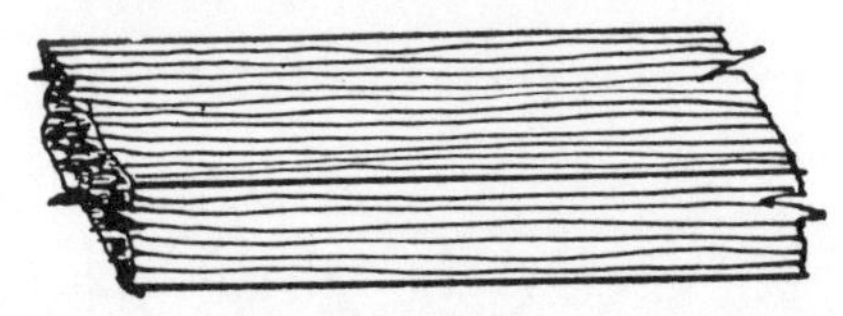

Fig. 416

Cross-Grained Lumber.—Fig. 414 shows a piece of cross-grained lumber having its grain running almost at a right angle with its sides or edges. Fig. 415 shows a piece of lumber partially cross-grained with the grain running, say, at about a 45-degree angle. Partially cross-grained lumber is also called, diagonally-grained lumber. The strength of the former is impaired much more than that of the latter. Lumber sawed from twisted tree trunks produces partially cross-grained lumber. Knots often cause both cross-grained lumber and diagonally-grained lumber. Fig. 416 shows a piece of straight-grained lumber.

Shakes.—Fig. 417 shows a heart shake. A star shake is shown by Fig. 418. Fig. 419 gives a cup shake, or wind shake, as it is also called. Some authorities do not make any distinction between a heart shake and a star shake, while others do.

Defects in Lumber.—Fig. 420 represents pitch pockets. Pitch pockets are filled with liquid or crystallized pitch. A rather long opening between the grain of the wood filled with pitch is called a pitch seam. A streak of solid pitch in lumber is called a pitch streak. When a portion of a piece of lumber is completely taken by pitch, it is called solid pitch.

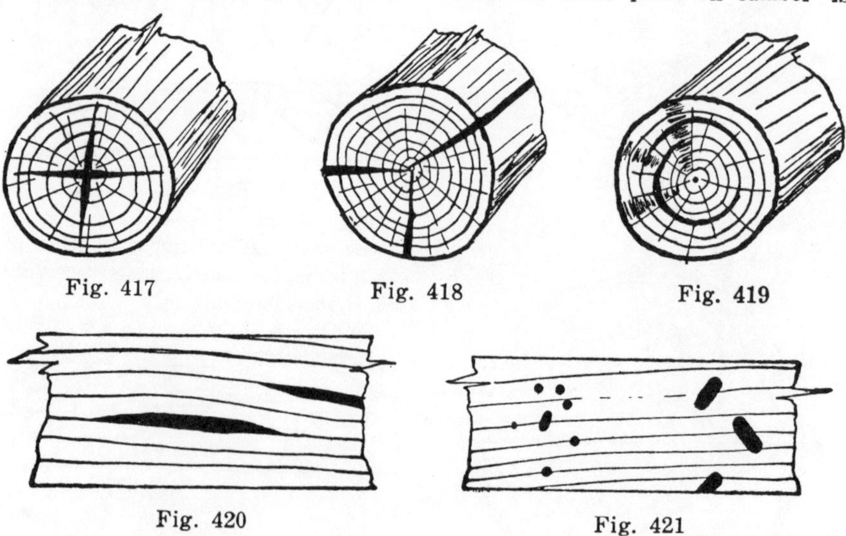

Fig. 417 Fig. 418 Fig. 419

Fig. 420 Fig. 421

Bark pockets, which impair wood just as pitch pockets do, are often referred to as pitch pockets.

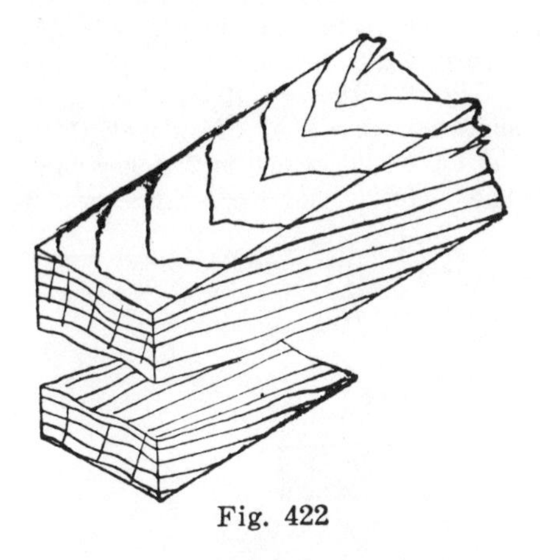

Fig. 422

Fig. 421 shows worm holes in lumber. To the left are small worm holes, while large worm holes are shown to the right. Besides worm holes in lumber, there are knot holes, picaroon

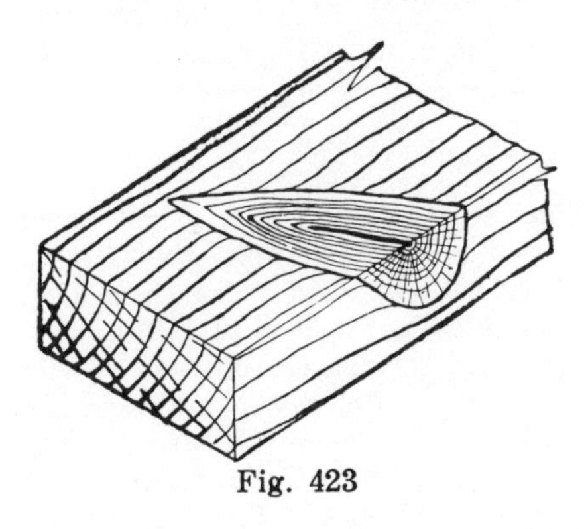

Fig. 423

holes, dog holes, nigger tears, bird pecks, insect holes, and rafting-pin holes, all of which constitute defects in lumber.

Diagonal Grain and Knots.—Fig. 422 shows a defect caused by diagonal grain. Diagonally-grained timbers should never be used for columns or for posts. In fact, such lumber should not be used where the timber is intended to support or carry some kind of a load.

Fig. 423 shows what is called a spike knot. Any other knot that is in some way similar to what is shown in the figure, would come under this classification. Round knots are more nearly circular in form. A good example of a round knot is shown by Fig. 424.

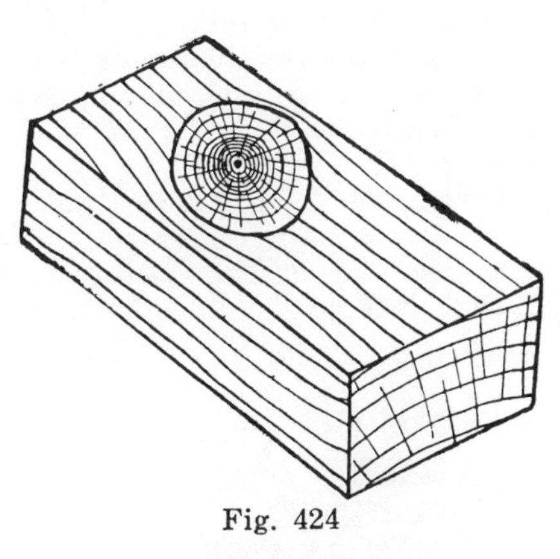

Fig. 424

A knot that is as hard as the wood around it is called a sound and tight knot. A knot not held firmly in position is a loose knot. A rotten knot is, as the name implies, rotten. A knot not over ½ inch in diameter is called a pin knot. One that has a ¾-inch diameter is a small knot. A standard knot is not over 1½ inches in diameter. All knots over 1½ inches in diameter are large knots.

LESSON 39

LUMBER

Lumber—I once saw a sign hanging in front of a carpenter's workshop which read:

"IF IT IS WOOD, WE CAN MAKE IT."

Then, of course, followed the name of the owner of the shop. Evidently what this man wanted to impress on the minds of those who read the sign was that his shop was a carpenter shop, and that he and his men could make anything that was made of lumber. Lumber could be defined as being the products sawed from logs, as boards, planks, scantlings, strips, shingles, lath, heavy timbers, and all kinds of finishing material, all kinds of moldings—in fact, if it is wood, it is lumber.

Yard Lumber.—Most of the lumber in our day is manufactured and sold according to standard sizes and grades. It is not our intention to do

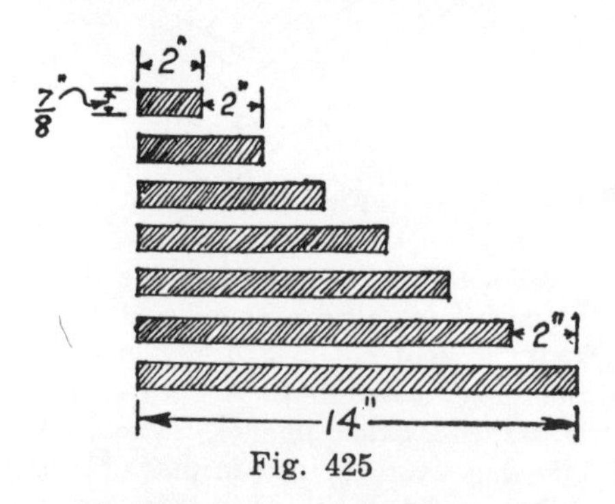

Fig. 425

any proverbial "hair-splitting," speaking in regard to the exact sizes into which lumber is sawed, but we will use the figures that are commonly used in practice. For instance, a board is sold as being 1 inch thick—it is commonly referred to as ⅞-inch stuff. Due to sawing and surfacing, its actual thickness is about 13/16 of an inch. Every carpenter knows that, as a rule, 2-inch lumber is only 1⅝ inches thick.

Fig. 425 shows end views of seven surfaced boards, ⅞ of an inch thick, the top one of which is 2 inches wide. The width, however, increases 2

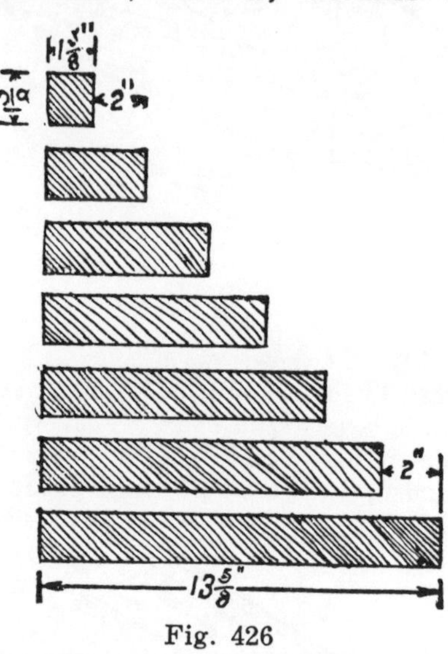

Fig. 426

inches on each board. The second board is 4 inches wide, the third is 6 inches wide, and so on down until the seventh, which is 14 inches wide. Common surfaced boards usually run full width, but not always. In either case, the buyer pays for full width. Widths of odd inches, as 1 inch, 3 inches, and 5 inches are seldom manufactured without a special order. This rule holds good in all unworked lumber. The lengths, too, run in even feet lengths, as 10, 12, 14, or some other even-feet lengths. Odd-feet lengths are not manufactured, as a rule.

Fig. 426 shows end views of seven pieces of 2-inch lumber. From top down each piece is increased two inches in width. This represents surfaced lumber, and the first, or top piece is only $1\frac{5}{8}$ inches by $1\frac{5}{8}$ inches, $\frac{3}{8}$ of an inch narrower than the size for which the buyer pays. The bottom piece, which represents a 2x14, is only $1\frac{5}{8}$ inches by $13\frac{5}{8}$ inches,

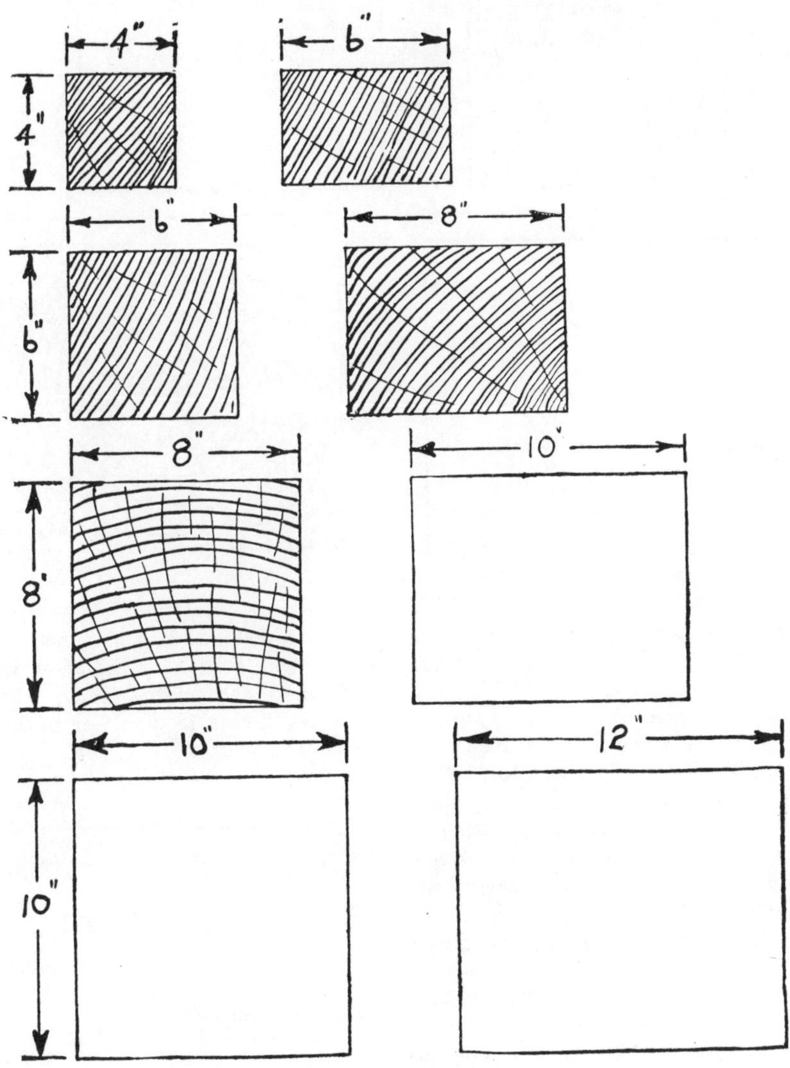

Fig. 427

although it is called a 2x2. The second piece is a 2x4, but measures only $1\frac{5}{8}$ inches by $3\frac{5}{8}$ inches. Each of the pieces, owing to sawing and surfacing, is $\frac{3}{8}$ of an inch thinner and scant. By "scant" is meant that because of shrinkage the wider pieces are often more than $\frac{3}{8}$ of an inch narrower than full width.

Structural Timber. — Unsurfaced,

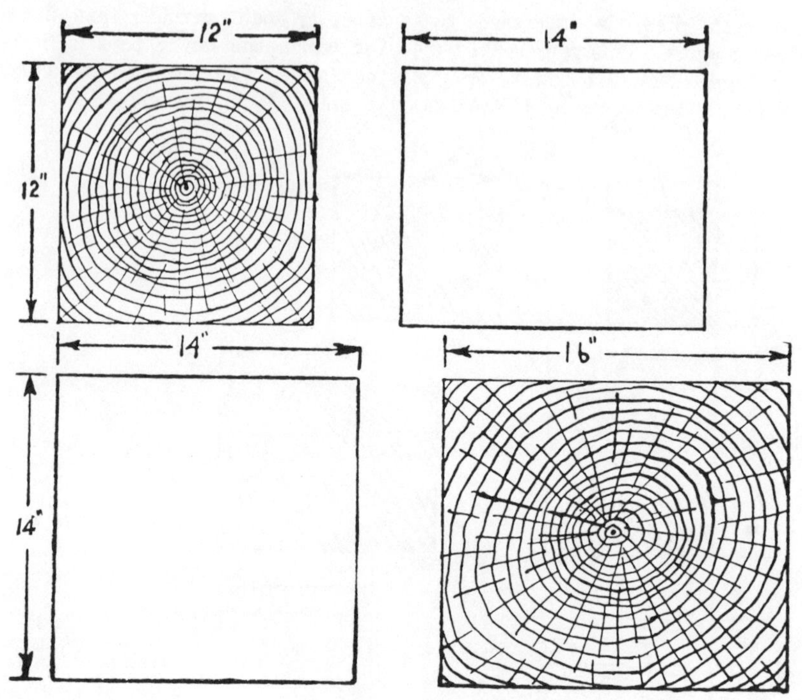

Fig. 428

or rough, lumber usually comes about full size. A rough 4x4 and a rough 4x6 are shown at the top of Fig. 427. These will measure in the rough, 4 inches by 4 inches, and 4 inches by 6 inches, nearly. Lumber that is less than 6 inches in thickness is called, yard lumber. Everything 6 inches in thickness and over is called, structural timber. The second, third, and fourth pairs of timbers, shown by Fig. 427, represent structural timbers. Fig. 428 shows four very heavy timbers. The two that are shaded show how they should be sawed, relative to the location of the heart. The dimensions being given on the drawings, no more needs to be said.

A few grading rules of select and common stock, in abbreviated form, are given here:

Select Common Boards. — These must be well manufactured and square-edged, and must not have knots over 1 inch in diameter in 4-inch to 6-inch widths, and no knots over 1½ inches in diameter in 8-inch to 12-inch widths.

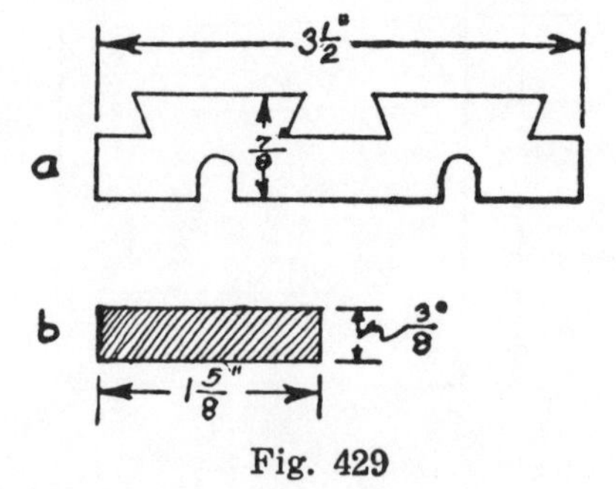

Fig. 429

No. 1 Common Boards.—Must not have knots over 1½ inches in diameter in 4-inch to 6-inch widths, and none over 2 inches in diameter in

8-inch to 10-inch widths. In 12-inch widths the knots must not be over 2½ inches in diameter, and not over 3 inches in diameter in widths wider than 12 inches.

No. 2 Common Boards.—This grade may have 2-inch tight knots in 4-inch to 6-inch widths, 2½-inch tight knots

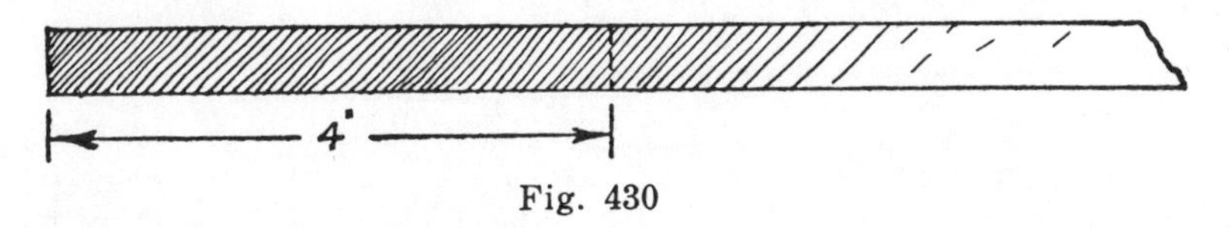

Fig. 430

in 8-inch widths, 3-inch knots in 10-inch widths, 4-inch knots in 12-inch widths, and in wider than 12 inches tight knots not over one-third of the width of the piece in which they are found.

No. 3 Common Boards.—This grade must not have knots over 1 inch in

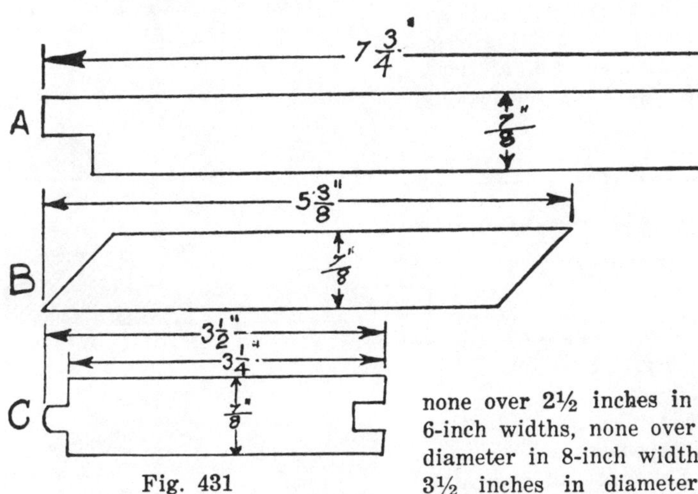

Fig. 431

diameter in 4-inch to 6-inch widths, none over 1⅜ inches in diameter in 8-inch widths, none over 1¾ inches in diameter in 10-inch widths, none over 2 inches in diameter in 12-inch widths, none over 2¼ inches in diameter in 14-inch widths, and no knots over 2½ inches in diameter in widths wider than 14 inches.

No. 1 Common Dimension.—This grade must not have knots over 1½ inches in diameter in 4-inch to 6-inch widths, none over 2 inches in diameter in 8-inch widths, none over 2½ inches in diameter in 10-inch widths, none over 3 inches in diameter in 12-inch widths, none over 3¼ inches in diameter in 14-inch widths, and no knots over 3½ inches in diameter in widths wider than 14 inches.

No. 2 Common Dimension.—This grade must not have knots over 2 inches in diameter in 4-inch widths, none over 2½ inches in diameter in 6-inch widths, none over 3 inches in diameter in 8-inch widths, none over 3½ inches in diameter in 10-inch widths, and none over 4 inches in diameter in 12-inch widths. In wider than 12-inch widths the knots must not be over one-third the width of the piece in which they are found.

Worked Lumber.—By worked lumber is meant any lumber that has been prepared for some special purpose by working it over with a matching machine, sticker, or molder. Finishing lumber, siding, flooring, and all kinds of moldings come under this head.

There was a time in the history of carpentry when a great deal of this material was worked over in the car-

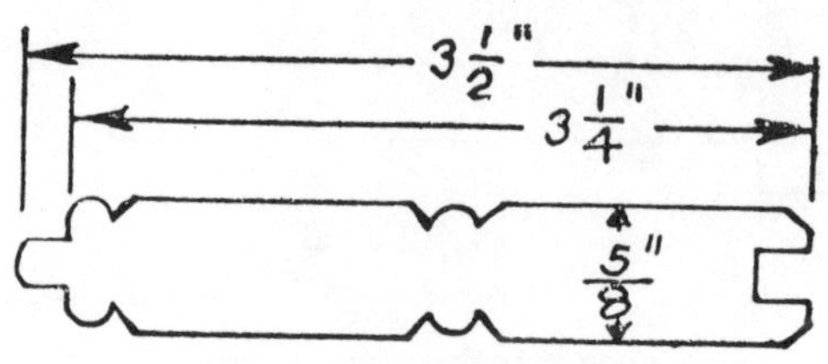

Fig. 432

penter shop, by the men who afterward used it in erecting the building. While now and then a little of this work is still done, as it were, on the job, the greater part of it is being done by the mills. In the days when carpenters had to make moldings, sash, and doors, they had to lug large tool chests with them full of tools. They had to have tools for making

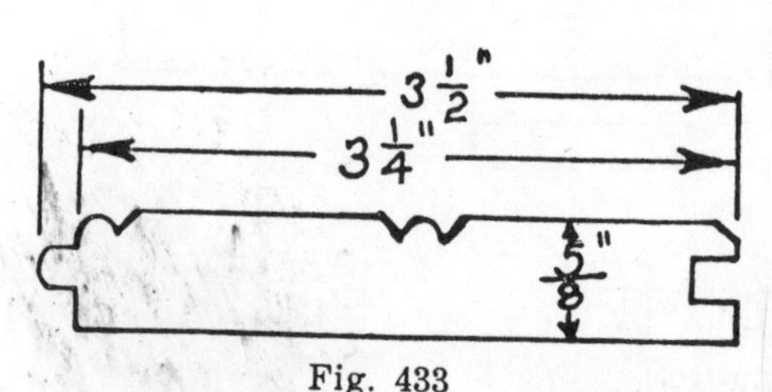

Fig. 433

moldings, beads, flooring, and other things that carpenters seldom have to do now. The work of this kind that the carpenter still does will soon disappear. And the large tool chests that our fathers carried from job to job will rot or turn into relics in barn lofts or in attics. The carpenter of today needs very few tools, compared with the tools needed by the old, old timers.

Fig. 429 shows, *a*, the end view of a narrow width of Byrkit (Burkett)

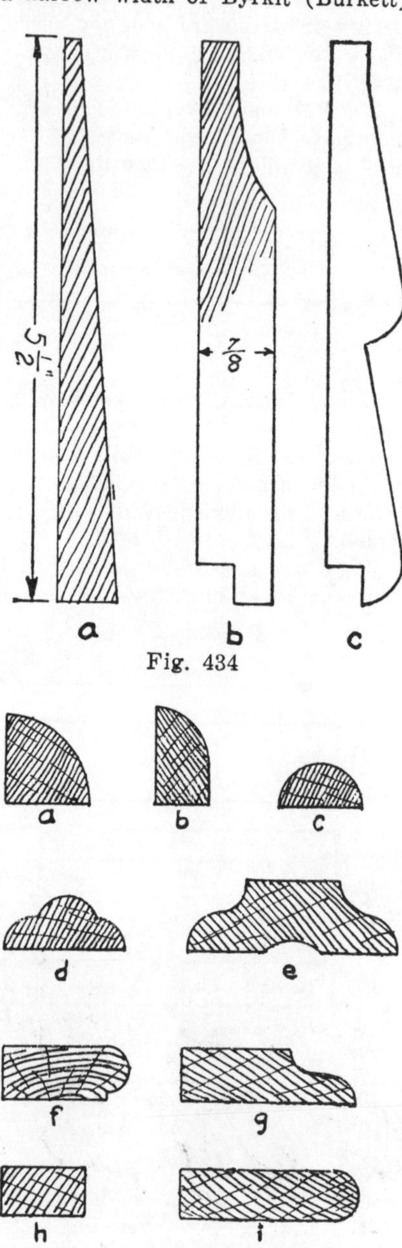

Fig. 434

Fig. 435

Lath, and at *b* is shown the end view of a common lath. Fig. 430 shows an

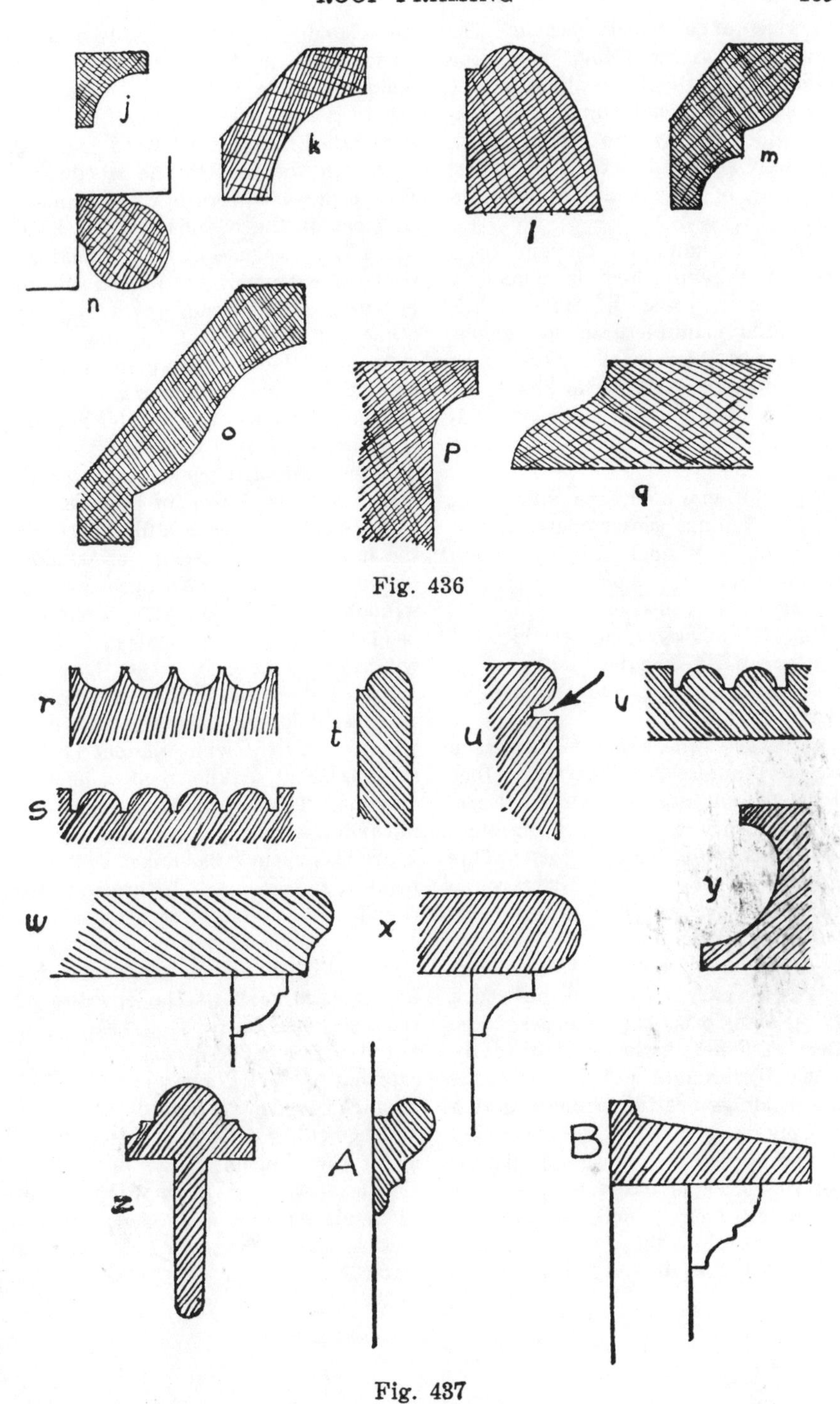

Fig. 436

Fig. 437

end view of a 4-inch shingle. The drawing shows the shingle extending beyond the 4 inches, which means that shingles do not run to any definite width. They are usually both narrower and wider than 4 inches, excepting, of course, dimension shingles.

Fig. 431 shows, *A*, the end of a piece of shiplap which is commonly known as a piece of 1x8 shiplap. Shiplap is manufactured in various widths, and there is also a variation in the thickness. At *b* is shown an end view of a piece of corn crib siding, and at *C* a piece of 3¼-inch flooring.

Fig. 432 shows an end view of a piece of ceiling, center matched and beaded on both sides: Fig. 433 gives an end view of a piece of ceiling matched and beaded on one side.

Fig. 434 shows, *a*, an end view of a piece of 5¼-inch lap siding; *b*, a piece of drop siding, and *c*, a piece of novelty siding.

Moldings.—The styles that pleased our forefathers do not altogether please us, and so, in order to please, designers bring to the foreground new styles from year to year. This principle is true in architecture as much as it is in other things. The architect studies and figures out new styles of architecture, and the carpenter is called upon to put such styles into practical realities. The effect of a new style of architecture is usually brought out by moldings. The moldings are the trimmings of a building.

Moldings are sold by the lineal foot, and the price is based on the piece of lumber from which the molding was worked, rather than on the amount of wood in the finished molding. Moldings that are made of soft wood give better satisfaction than moldings made of hard wood, excepting in cases where the finish requires a certain kind of hard wood molding to match the rest of the woodwork. Where the woodwork is to be painted, as most of the outside woodwork is, there is no reason why moldings made of extremely hard wood should be used. White pine, cypress, and other soft woods that will last well, exposed to the weather, give excellent results for outside finishing.

Fig. 435 shows, *a*, a quarter round; *b*, a base shoe; *c*, a half round screen molding, and *d*, a clover leaf screen mold. At *e* is shown an ogee batten, at *f* a molding used as a fillet between the head and side casings of window and door openings. An ogee door or window stop is shown at *g*, a parting bead at *h*, and a round-edged door or window stop at *i*.

Fig. 436, *j* and *k*, show two styles of cove moldings. The one shown at *j* is a solid cove, while the one shown at *k* is what is known as a sprung molding. The bed molding shown at *m*, and the crown molding shown at *o*, are also sprung moldings. A thumb molding is shown at *l*, a three-quarter round at *n*, a congee mold at *p*, and an ogee (O G) at *q*.

Fig. 437, *r*, shows what is called fluting, and *s* shows the opposite or reeding. The half rounds shown at *t* and *u* are called beads. The indicator at *u* points out what is called a quirk. A double bead is shown at *V*. A molded nosing is shown at *w*, and a round nosing at *x*. A scotia is shown at *y*, and at *z* we have an astragal. At *A* is a common form of picture molding, and at *B* is shown a water table.

INDEX